IMAGES
of America

CITY OF INDUSTRY

On the Cover: The five men in this photograph were instrumental in the conception and early administration of the City of Industry. The photograph was taken in 1958 in front of the first city hall, during its construction. The men are, from left to right, John Ferrero (mayor), Robert Rope, (city manager), James Stafford (business owner), Vicente Perez (business owner), and Darius Johnson (council member). (Courtesy of the Perez family.)

IMAGES
of America

CITY OF INDUSTRY

Jeff L. Parriott for the City of Industry

ARCADIA
PUBLISHING

ISBN 978-1-4671-0778-5

Published by Arcadia Publishing
Charleston, South Carolina

Library of Congress Control Number: 2021948249

For all general information, please contact Arcadia Publishing:
Telephone 843-853-2070
Fax 843-853-0044
E-mail sales@arcadiapublishing.com
For customer service and orders:
Toll-Free 1-888-313-2665

Visit us on the Internet at www.arcadiapublishing.com

Contents

ACKNOWLEDGMENTS

A very special thanks to the City of Industry's city council and assistant city manager/public affairs manager Sam Pedroza for having the vision to pictorially document the history of the city and in the supplying of photographs, news clippings, and many other sources of information.

Paul Spitzzeri, director of the Workman and Temple Family Homestead Museum, was especially helpful with early area photographs from the museum's collection, being a vital resource as a regional history authority, and proofreading of the caption drafts.

A thank-you goes to my friend Vince Perez for sharing his knowledge and perspectives as a member of our team and to his family for providing numerous photographs.

Great appreciation goes out to the Industry Business Council and all the past chambers of commerce that archived decades of newsletters related to the city, its businesses, and residents. To the city's civil engineer, CNC Engineering, and Clem Calvillo's knowledge that spans decades of projects that built the city, I greatly appreciate their assistance.

And thanks to those who shared their stories and the stories of others with many of us through the years, but who are no longer here, among them Mayor Dave Perez and Phyllis Tucker.

Some images in this volume appear courtesy of the City of Industry (COI) and the Workman and Temple Family Homestead Museum (WTFHM).

Introduction

During the post–World War II era in this region east of the burgeoning metropolis of the city of Los Angeles, the stage was set for a great transformation of the land and its inhabitants. It had been some 70 years since the first transcontinental railroad came to the eastern San Gabriel Valley, bringing immigrants and land speculators to the ranchos and small towns. The rail lines of the Southern Pacific and its rival the Union Pacific created a swath of land between the two that sparked a battle for its possession.

The Los Angeles Regional Planning Commission had been eyeing this elongated U-shaped ribbon of unincorporated farm and ranch land for use as an industrial zone since the 1920s. It was not until 1954, when the commission followed-up with a comprehensive plan that consisted of 3,500 acres stretching from El Monte in the west to Pomona in the east. The contenders in the battle for the land included the incorporated city of West Covina, the unincorporated towns of Walnut and Puente, and landowners within the zone, including the Southern Pacific Railroad. Neither West Covina nor any of the towns were able to garner the support of the railroad and landowners to incorporate into their boundaries.

Visualizing an industrial municipality that would avoid (as much as possible) interference with a residential community, the Southern Pacific, along with individual property owners and businesspeople, set out to incorporate a city strictly for business. In a formal statement by the chairman of the Committee for the Incorporation of Industry, California, Edward Lustgarten, owner of the 378-acre Suzy Q Ranch near the intersection of Colima Road and Jellick Avenue, said the following: "Perhaps the largest concentration of industry in the State of California will exist in the city we propose, and, while we are more than satisfied with the administration of the county government which we now enjoy, we wish to remain the distinct area that lies within our boundaries." Lustgarten added, "Our sole aim is to attract additional industry, and it is hoped that we can maintain complete freedom from the tax burdens which confront other cities."

The requirements for incorporation were that the petitioners must represent 25 percent of the property owners and 25 percent of the assessed valuation of the properties within the proposed boundaries. Among the large landowners of the proposed city was the estate of Clara Baldwin Stocker, daughter of Elias "Lucky" Baldwin, the man who, in the late 1870s, acquired the William Workman portion of Rancho La Puente after Workman defaulted on a loan to his bank. The Stocker holdings were in the area of Valley and Hacienda Boulevards and were a part of the one-third of her estate left for the sole purpose of benefitting a home for elderly women, which today is a skilled nursing facility.

The Sunkist Corporation was another of the large landholders whose involvement in the citrus industry was dwindling in Los Angeles and Orange Counties due to the increase of suburbanization and the movement of the groves north to Ventura and Santa Barbara Counties.

On December 4, 1956, residents within the proposed city voted 118-24 in favor of incorporation. The certification was stalled by a lawsuit that was dismissed, and it was finalized on June 18, 1957, with the first city council meeting quickly arranged for two days later in a conference room at Utility Trailer Company.

From that moment, a whirlwind of development commenced that likely exceeded the wishes expressed by Lustgarten of attracting industry into this new city. The city council had the foresight to adopt the Lakewood Plan of contracting services from the county and others in order to reduce operational and maintenance costs and to take advantage of the vast resources of one of the nation's largest counties. With the consistency of the long-serving team of Mayor John Ferrero, council members Charlie Rowland and Darius Johnson, city manager Bob Rope, and others, the city had a strong chance to thrive.

Attracted by inexpensive land costs and an absence of city property and business taxes, companies large and small flocked to the city, either relocating from high-taxing cities or expanding from as far away as the East Coast.

Using the city's redevelopment agency as an effective tool to finance and construct the necessary infrastructure of storm drains, sewers, streets, and lighting, the former farmlands were transformed into ready-made sites for the industrial interests that followed. Soon, the area in which the county had planned to have a hefty employment base for its residents was bustling with a large number of good-paying manufacturing jobs.

The city rounded out its portfolio in the late 1970s and early 1980s with recreational and historical/cultural projects for all to enjoy. Currently, the city is undertaking efforts with the Industry Business Council to bring high-quality manufacturing jobs back within its bounds, as well as offering a foreign-trade zone to help companies minimize the expense of US Customs duties. And while the focus of the city has and will be the providing of jobs to residents of local communities, it has not lost sight of their needs beyond the workplace. Under the leadership of its first female mayor, Cory Moss, the city has taken an active role in the local food bank and autism awareness training for law enforcement and education officials. Many other city-supported endeavors are featured within the pages of this book.

Our hope is that the reader will have gained insight into what it was like to conceive, foster, and grow a unique industrial city through its first six decades. And, while there have been those who have taken selfish advantage of their nonelected/appointed positions around the city's business, no elected or staff persons have ever been charged with any wrongdoing.

While researching the people involved with the early administration of the city, I found that many served our country during World War II and would like to acknowledge them here: city manager Robert S. Rope, US Navy and Naval Reserve; city attorney Glenn R. Watson, US Navy; city attorney Robert G. Beverly, US Marine Corps; planning commissioner Alfred Benton, Army Air Corps (two-time Distinguished Flying Cross and Silver Star); and John Radecki, US Army.

As this is an abbreviated history of the City of Industry and the thousand words that a picture is worth is not quite reached, we hope that an appreciation is realized for the people and efforts needed in the founding, growing, and sustaining of a unique city.

One

A Century of Ranchos, Railroads, and Revision

William Workman (1799–1876) was originally from Clifton, England, and is seen here in 1852 with his daughter Antonia Margarita. Workman came to the area that was to become the City of Industry via Taos, New Mexico, and Los Angeles in 1841, and co-owned the Rancho La Puente. He was married to Nicolasa Urioste, a native of Taos. (WTFHM.)

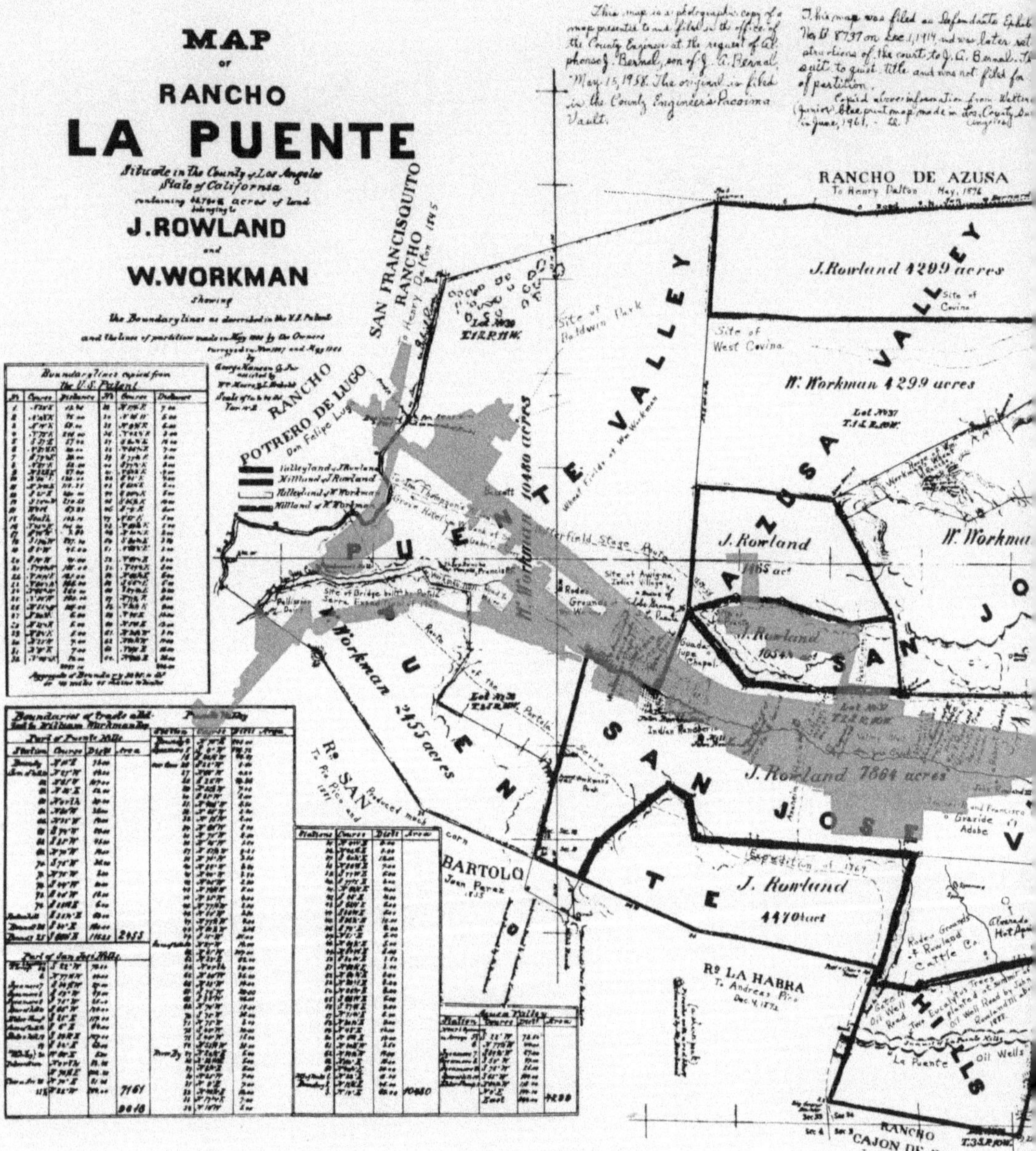

MAP
OF
RANCHO
LA PUENTE
Situate in the County of Los Angeles
State of California
J. ROWLAND
and
W. WORKMAN
Showing
RANCHO
POTRERO DE LUGO
Don Felipe Lugo
SAN FRANCISQUITO
RANCHO
RANCHO DE AZUSA
To Henry Dalton May, 1876
J. Rowland 4299 acres
W. Workman 4299 acres
Site of Covina
Site of West Covina
Site of Baldwin Park
PUENTE VALLEY
AZUSA VALLEY
SAN JOSE VALLEY
PUENTE HILLS
SAN JOSE HILLS
W. Workman 10480 acres
W. Workman 2455 acres
J. Rowland
1465 acres
J. Rowland 7864 acres
J. Rowland
4470 acres
Butterfield Stage Route
Indian Rancheria
BARTOLO
Juan Perez
Rº LA HABRA
To Andreas Pico
Dec 4, 1872
Rº SAN
RANCHO
CAJON DE SAN JUAN
To Juan B. Ontiveros
Pacifico 1877
La Puente Oil Wells
Rodeo Grounds of Rowland Cattle Co.
Expedition of 1769
Alvarado Hot Springs
This map is a photographic copy of a map presented to and filed in the office of the County Engineer at the request of Alphonse J. Bernal, son of J. A. Bernal, May 15, 1958. The original is filed in the County Engineer's Pacoima Vault.
This map was filed as Defendants Exhibit No. D 8737 on Dec 1, 1914 and was later set by instructions of the court to J. A. Bernal ... suit to quiet title and was not filed for of partition.
Copied above information from Walter (Junior) blue print map made in Los County Survey in June, 1961.

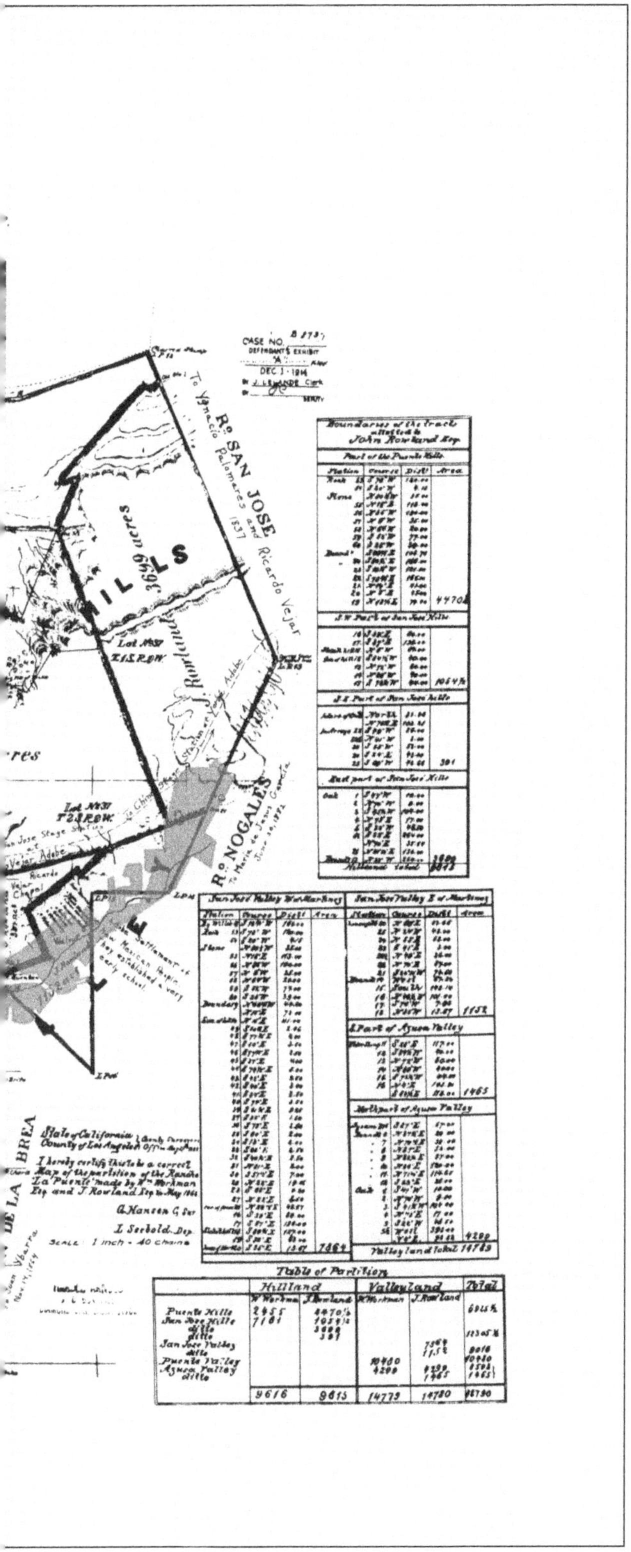

This 1868 map of the Rancho La Puente and the City of Industry shows how they are inextricably linked with their east-to-west length being nearly identical and encompassing what was land granted to both William Workman and John Rowland. The natural environment of the Rancho La Puente consisted of fertile farmland, water sources such as the San Gabriel River and San Jose Creek, abundant livestock pasturage, and a Mediterranean-like climate. Among the inhabitants of the area were the native people, Spanish settlers, and missionaries. Infrastructure in the early years consisted of the Butterfield Stage Route, roads such as the Colorado or County Road (Valley Boulevard), and Anaheim to Puente Road, now roughly Harbor Boulevard, Fullerton Road, and Azusa Avenue, traveling north and south. (WTFHM.)

Pictured about 1860, John Albert Rowland was William Workman's co-grantee of the 48,000-acre Rancho La Puente after the two led a group of 65 from New Mexico to Los Angeles. The land was formerly owned by the San Gabriel Mission before secularization. Rowland lived on his ranch until his death in 1873, having several children from two marriages. (WTFHM.)

This native Indian kizh (house) on the Rancho Santa Anita, where the original village was called Aleupkinga, was likely photographed about 1870 by William M. Godfrey. There were two probable native sites on the Rancho La Puente, one of them in the area of William Workman's El Campo Santo cemetery and the other in the present area of La Puente City Park. (WTFHM.)

William Workman built an adobe home on his half of the rancho and was successful in the raising of grain, wine grapes, and cattle ranching and, later, added a winery and distillery. Workman was soon able to expand the home to what is seen here around 1872; it still stands today. (WTFHM.)

William Workman built the St. Nicholas Chapel, shown here around 1890, within El Campo Santo; it is one of the oldest private cemeteries in Southern California. The artist Henry Miller made drawings for Workman and stated that the chapel was to be built for the benefit of the indigenous people who worked for Workman, earning 50¢ a day. (WTFHM.)

With the completion of the Southern Pacific Railroad line to the Puente Station in 1874, the farmers, ranchers, and retailers of the Rancho La Puente area had access to a new distribution and supply source. The c. 1890s photograph above shows the Bassett train station, which was situated on the line at the present-day location of Valley Boulevard and Temple Avenue in the City of Industry. The Atchison, Topeka & Santa Fe Railroad was completed north of Puente in 1885. This was the second transcontinental rail line to serve the area of the old rancho, and this arrangement would be a boon many decades later to the development of the City of Industry. The photograph below was taken near Puente in 1897. (Both, WTFHM.)

In stark contrast to William Workman's successful farming and ranching endeavors, his foray into the banking business with his son-in-law Francis Pliny Fisk Temple was a catastrophic failure. The Hellman, Temple and Company bank opened in the Pico Building in Los Angeles in September 1868, during a period of great growth in the city. After two years, Temple and Workman's partner Isaias W. Hellman broke from the alliance because of a difference in lending philosophies between he and Temple. In November 1871, the Temple and Workman Bank opened in the Temple Block, photographed by Henry T. Payne around 1872. Hellman meanwhile formed Farmers and Merchants Bank with former governor John G. Downey. When the economy went bust in 1875, poor management and inadequate cash reserves left the Temple and Workman bank vulnerable to nervous depositors. A loan was secured from Elias J. "Lucky" Baldwin with Workman's portion of the rancho used as collateral. The loaned money did not last, and the bank closed in January 1876. (WTFHM.)

Shown here around 1890 are laborers harvesting an orange grove in the San Gabriel Valley. The North Whittier Heights Citrus Association packing plant, in what was named Hillgrove, was located on the San Pedro, Los Angeles & Salt Lake Railway, at what is today the intersection of Clark and Ninth Avenues in the City of Industry. (WTFHM.)

Five months after the collapse of the Workman and Temple bank in 1876, William Workman committed suicide at his home on the Rancho La Puente. "Lucky" Baldwin foreclosed on the property, and it was subdivided and sold off with the lot that included the Workman Home whittled down to 75 acres. This c. 1900 photograph depicts hay being harvesting on the ranch when it was owned by Frederick Smith. (WTFHM.)

Many of the local families were very tight-knit, as they often joined in ventures of business, farming, and intermarriage. This c. 1900 photograph shows members of several local families, including the Rowlands, Yorbas, Sanchezes, Pilarios, Vejars, Grazides, and Dunphys. The Yorba family has been in the area since the early 1770s and owned ranchos in modern Orange County. Ricardo Vejar was the co-grantee of Rancho San Jose, in the Pomona area, with Ygnacio Palomares. A French Basque sheepherder, Francisco Grazide had a large ranch in what is now Rowland Heights and was married to Isabel Rowland, granddaughter of John. Included in this photograph are young siblings Charles J. and Elvira Rowland; he was elected to the first council of the City of Industry, and she was elected as the first city clerk. The occasion and location for this photograph are unknown; however, it may have been one of the Rowland family's many barbecues over the years on their portion of Rancho La Puente. (WTFHM.)

The roads of this era were, of course, rudimentary, as depicted here in an unattributed c. 1900s photograph, titled "Through The Willows—San Gabriel River." The road is unidentified and is a far cry from any of today's roadways that cross the San Gabriel River, such as Valley Boulevard. Valley Boulevard has been a main east-to-west thoroughfare for hundreds of years and was known in its early years as the Colorado Road because it stretched all of the way to the Colorado River at Yuma. Other monikers included Puente–San Jose Road and Pomona Boulevard. Its route is shown on maps created in the 18th century, and today, it covers much of its original course, save for some realignments to accommodate later developments. (WTFHM.)

Little is known about Frederick Smith's ownership of the Workman Homestead before he sold the property in 1903 to Lafayette F. Lewis of Anaheim. Around this time, the St. Nicholas' Chapel had apparently burned down, and Lewis proceeded to desecrate the El Campo Santo cemetery to clear more land for grazing. This photograph of El Campo Santo shows some of the cemetery walls destroyed and the chapel gone in 1906. William Workman and Nicolasa Urioste's grandson Walter P. Temple filed a lawsuit against Lewis and obtained a court judgment of $5,000 to repair the cemetery walls. Instead of paying the fine, Lewis sold the Workman Homestead in 1907 to Eugene Bassett, a Pasadena-based real estate agent, and his son-in-law Thurston H. Pratt. (WTFHM.)

Thurston Pratt is shown above with a child in a large wheelbarrow and below working on a farm implement with an unidentified man in 1912 on the Workman Homestead. Pratt and his family lived in the Workman House and actively farmed the homestead land, growing hay, grain, and oranges. They were known for hosting picnics near the San Jose Creek, as well as parties, fundraisers, and barbecues on the property. Pratt was an involved member of the Puente community and was chosen to oversee a school bond election in 1909 and elected to a committee of the newly organized community improvement society in 1911. (Both, WTFHM.)

In 1860, it was recorded that William Workman had 50 acres of vines planted and 5,000 gallons of wine on hand. Workman built his own winery, which consisted of three buildings, and among the necessary implements for the winemaking operation were barrels for aging. This 1912 photograph, taken when the homestead was owned by Bassett and Pratt, shows two women sitting atop one of the old wine barrels. (WTFHM.)

Pictured here in 1912, Thurston Pratt is cutting what may well be his crop of oat hay or other grain, as a young Norman King and his mother, Daisy King, observe. All of these folks migrated here as part of the population explosion between 1900 and 1920 when the county went from 170,000 residents to 936,000. (WTFHM.)

The first commercial orange grove in the area was planted by William Wolfskill in Los Angeles in 1841. The market for the local oranges expanded nationwide in the 1890s with the improved refrigerated railroad boxcars. Seen here in April 1912 with some of that year's harvest are, from left to right, Daisy King, Frank Harpine (ranch employee), Eugene Bassett, and, sitting on the ground, Norman King. The oranges featured in this photograph were likely from the ranch's trees, including the one shown behind the group, located near the Workman House. It was reported in this same year that Pratt and Bassett had a crop of olives that were sold to the J.C. Grogan Company. As would happen with the local walnut farms, the orange groves gave way to the suburbanization of the valley after World War II. (WTFHM.)

Riding on the top of a fully loaded hay wagon of Pratt and Bassett are Daisy King (left) and a woman only identified as Mabelle. The price for the oat hay advertised that year as "gilt edge" was $25 per ton, while the previous year fetched only $15. The buildings in the background are the winery structures built by William Workman in the 1860s. Below is an unidentified group of women on the Workman Homestead playing croquet. (Both, WTFHM.)

During this period of great agricultural activity in the area, there was a need for labor that was largely filled by minority groups. In this region, it was usually people of Japanese and Mexican descent who were employed at farms. These two 1912 photographs show scenes from campsites near the Workman Homestead. Pictured above is a man at a large pot or washbasin, located next to what appears to be a vehicle. Pictured below are women and children standing next to a building with hand tools leaning against it. (Both, WTFHM.)

In 1913, the Puente Rancho Packing Company formed and operated on the Workman Homestead property. In advertisements for stock purchasing, the plant's location between the two railroad lines and "On the good roads" was touted. The winery structures and other buildings on the property were repurposed for meat packing and canning of fruits, vegetables, relishes, and sauces. The company apparently lasted until shortly before Walter P. Temple, the grandson of William Workman, bought the Homestead from Pratt and Bassett in November 1917 for $40,000. (*Los Angeles Times*, 16 March 1913; *Los Angeles Express*, 10 May 1913.)

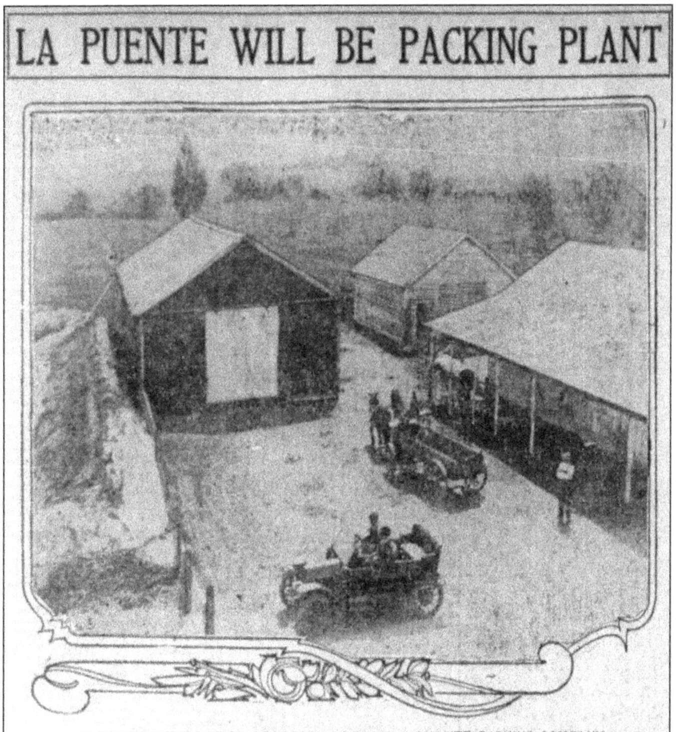

Oil was first discovered in the area during the early part of the 1880s by William Richard Rowland, son of John and Encarnacion Rowland. William went on to be highly successful in this endeavor, and his Puente Oil Company's 70 oil wells merged with 3 other oil interests in 1912. These two c. 1917 photographs show scenes from Walter Temple's oil leases in the Montebello Hills, just a short distance west of the City of Industry's westernmost boundary. The oil was discovered by Temple's nine-year-old son, Thomas, while he was playing in the hills above their home. Temple, by happenstance, was following in the footsteps of his father, F.P.F. Temple, who was among the first oil explorers of the greater Los Angeles area in the 1870s near today's Santa Clarita. (Both, WTFHM.)

Through his newfound wealth, Walter Paul Temple was able to purchase the Workman Homestead that his grandfather established over seven decades earlier. Shown here in two portraits in 1919 by George Steckel are Walter and his wife, Laura González, and the couple with their children, from left to right, Walter Jr., Agnes, Laura, Edgar, Walter Sr., and Thomas. All three boys were students of the Pasadena Military Academy, hence the uniforms. (Both, WTFHM.)

In this c. 1921 aerial view by Spence Air Photos is the Workman and Temple Homestead. Walter P. Temple had acquired the property four years earlier. At the center of the photograph is the house built by his grandparents, William and Nicolasa Workman, along with three winery buildings, a swimming pool, and tennis court. The grove to the left of the house are the walnut trees that were grown by all three generations of the two families. To the left of that is the meandering course of San Jose Creek as it heads to the San Gabriel River beyond the top of this photograph. The straight line farther left is the Union Pacific Railroad, formerly the Salt Lake line, and at the bottom left is El Campo Santo with the newly constructed Walter P. Temple Memorial Mausoleum. The road that leads from that point upwards was named Evergreen Lane, and it terminates at what is now Turnbull Canyon Road, formerly Tenth Avenue. (WTFHM.)

With the proceeds from their oil lease, Walter and his wife, Laura, began construction of a new adobe home, La Casa Nueva. Depicted here around 1927, the house is of the Spanish Colonial Revival architectural style and is located next to the Workman House, which can be seen to the left. Unfortunately, Laura passed away in 1922, just as construction began. (WTFHM.)

The Temples hired master builder Pablo Urzua and his crew from Guadalajara, Jalisco, Mexico, to make the adobe bricks from the Homestead's own soil. Shown here in about 1924 is the adobe brickmaking operation with the pit dug for the brick's clay in the foreground. The walnut trees surrounding the operation were another resource for the brickmaking, as branches were used to stoke the kiln's flame. (WTFHM.)

Besides building La Casa Nueva, Walter and Laura Temple made improvements throughout the Homestead property not only for their family but also for the community at large. One of the three winery buildings was transformed into an auditorium with a stage, film projection equipment, ping-pong, and pool tables, another was turned into a large cafeteria, and the last became a nine-car garage. The c. 1920 photograph above was taken on the occasion of a film screening that had an attendance of nearly 100 family and friends. The courtyard of La Casa Nueva is the setting of the other photograph in spring of 1928. This costumed group is composed of members of the St. Joseph's Church Altar Society from Puente. Among the attendees are representatives of the Romero, Temple, and Didier families. (Both, WTFHM.)

Among the numerous farmers in the area during the early part of the 20th century was Giuseppe (Joseph) Alberto Ferrero. In 1886, he immigrated from Italy and purchased farm property in about 1903 between the two local railroad lines, in what is now the City of Industry. Ferrero and his wife, Domenica, were well known in the community, and in April 1918, they hosted a barbecue for their son George, who was leaving for military training camp. He also used the occasion, as it was reported by the *Los Angeles Times*, for several Italian speakers to encourage the attendees to purchase Liberty Bonds, of which, $27,000 worth were sold. The other connection Fererro has to the City of Industry is that his granddaughter Verabelle Zabaldano would be elected as the city's first treasurer in 1957. Shown here around 1910 is a group of men inside a garage that was on the Ferrero farm. (WTFHM.)

Before his death in 1909, Elias J. "Lucky" Baldwin planted many acres of walnut trees on the land that was previously William Workman's portion of the Rancho La Puente. These groves thrived, and subsequent owners of the subdivided rancho formed the La Puente Valley Walnut Growers Association in 1912. Among the original members were William R. Rowland, Anna Hurst and her son Charles (whose property is a West Covina historic site), Mrs. F.S. Stimson, and Louis Didier. Pictured here is an unidentified grove near Puente in 1922 and the original 1920s packinghouse. The 5,000-square-foot packinghouse was built in 1913 on the north side of the Southern Pacific Railroad in Puente and stills stands today at the same location. (Both, WTFHM.)

As the acreage of mature walnut groves producing marketable nuts boomed, a larger processing plant was needed to handle the expanding industry. The world's largest walnut packing plant was built in what is now City of Industry, shown here a short time after its construction in the 1920s. The location was on the north side of the same rail line as the first plant but to the west of Hacienda Boulevard. The packinghouse was over 50,000 square feet, used a 160-horsepower load of electricity to run the most modern equipment, and had a 150,000-gallon water reservoir for fire suppression. A large number of laborers, mostly women, were hired to work in the processing of the nuts. Peak production was reached in 1927 when there were 7,000 acres producing about 45,000 tons. By 1950, there were just 425 acres in production. This was because the Great Depression lowered prices, widespread infestation slowed production, and suburban development shrank the agricultural acreage. (WTFHM.)

Cecilia Olivares is shown here around 1920 in the driveway of her parents, Luis and Rosenda's, house. Situated on the old Anaheim Puente Road that is now Azusa Avenue, the property included a walnut orchard, hayfield, a barn, and shop typical of the area. In 1842, Cecilia's maternal great-great-grandfather, Mariano Chavez, came to the area, as Workman and Rowland did, by way of the Old Spanish Trail. Cecilia continued to live in the house after her marriage to eventual City of Industry council member, Samuel J. Parriott, and did all but the driving of the trucks in the operation of his trucking business. Shortly after the city incorporated, Cecilia and her sister Luisa Vejar filed for the second annexation of property to be added to the new municipality. In the early 1960s, the state bought the property via eminent domain in order to construct State Route 60, the Pomona Freeway. (Courtesy of the Parriott family.)

Vicente Perez, shown here at his International Products Company store around 1925, was a naturalized citizen from Spain who immigrated to Los Angeles via New Orleans, Louisiana, in 1919. Perez opened his store on Valley Boulevard, just east of what is now Turnbull Canyon Road. He was very involved in the community and would be an early proponent of the City of Industry. Vicente would go on to establish the City of Industry Disposal Company and operate it, and other ventures, with his three sons, one of whom, David, would become the fourth mayor of the city. (Both, courtesy of the Perez family.)

Ample Space for Industries as Los Angeles Grows

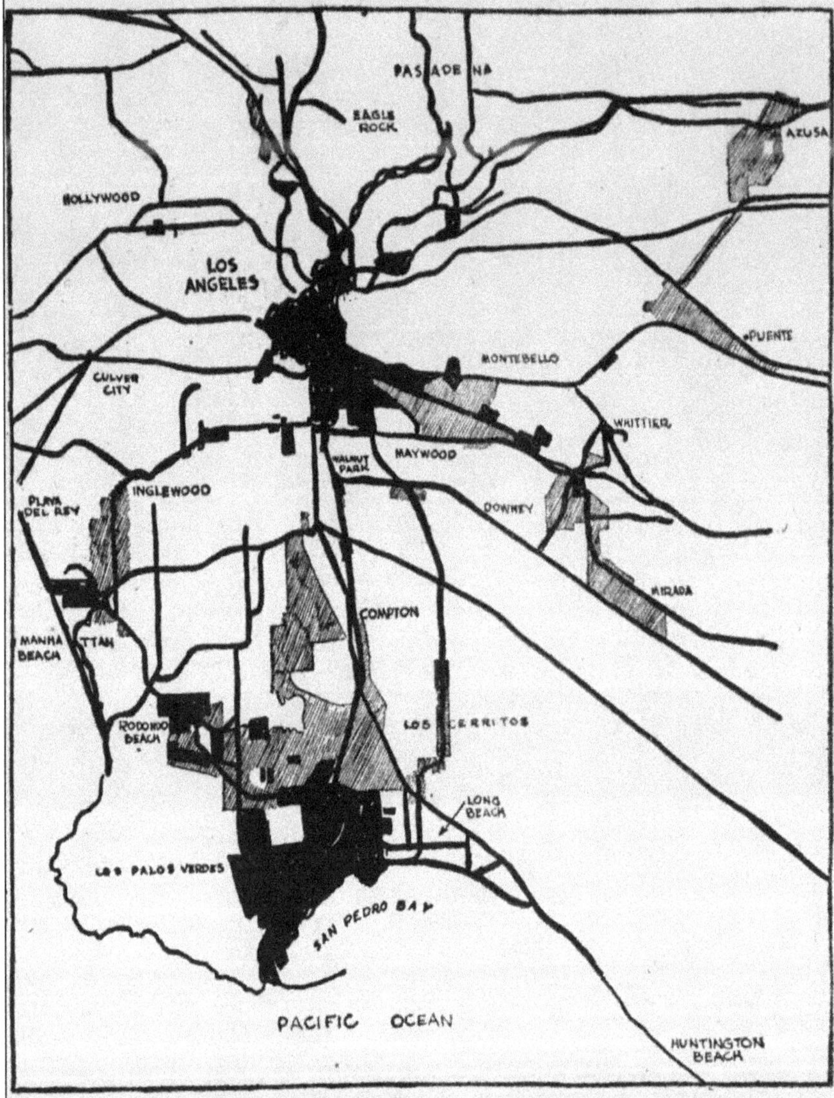

Regional Planning Commission Specifies Areas

Black areas shown on map are the present industrial sites. The shaded portions are the areas tentatively designated by the commission in the industrial site expansion program. These expansion areas are being handled unit by unit in definitely setting aside permanent industrial locations. Other sections segregated as future industrial sites by the planners and not shown on the map include small areas at Chatsworth, San Fernando and Pomona.

Los Angeles City was booming in matters of population and industry in the 1920s. As part of the Los Angeles Regional Planning Commission's strategy to keep ahead of the industrial growth, a study of 18 months produced this approach to the future needs of land use. William J. Fox was the chief engineer of the commission that stated that the prime benefit of a solid industrial property is its proximity to ample transportation, such as highways and railroads. On the upper-right portion of the map, in what was then Puente and now City of Industry, is the commission's inclusion of a planned industrial area. Fox noted that residential uses have been, and could continue to be, a hindrance to the most effective placement of industrial interests if better planning was not utilized. (*Los Angeles Times*, 18 March 1928.)

This 1930s aerial photograph shows the C.C. Stafford Feed Mill and the La Puente Valley Walnut Growers Association packing plant. In the foreground of the buildings are the Southern Pacific Railroad line, Valley Boulevard, and an empty field. Behind them are Puente High School, St. Joseph's Church, and various buildings, fields and groves. The location today is partially occupied by the Los Angeles County Sheriff's Department's Industry Sheriff's Station, helipad, and temporary buildings at Stafford Street and Hudson Avenue in the City of Industry. Josiah Whitcomb Hudson, for who the avenue was named after, was born in New York about 1843, served in the Iowa Infantry during the Civil War, and came to this area some time before 1880. In 1879, he married John Rowland's daughter Victoria and farmed in the area until about the time of his death in 1913. Hudson's name was also commemorated on the first school and school district in Puente. (COI.)

During the 1930s, the agriculture industry in Los Angeles County was flourishing. With occasional setbacks caused by disease and higher feed costs due to the drought in the Midwest and locally, the livestock trade was, for the most part, prosperous. In this aerial view from that period is an unidentified stockyard along Valley Boulevard and the Southern Pacific Railroad. The San Jose Creek winds around its southern border and separates it from more farmland. At the top and center is the curving road of Hatcher Avenue, turning west into Chestnut Street. It was reported in 1929 that $30 million worth of livestock from all over the region passed through the Union Stockyard in Los Angeles. In 1939, the figure for Los Angeles County alone was over $32 million, with beef comprising almost $12 million. (Courtesy of the Perez family.)

When this Spence Air photograph was taken in 1932, the Joseph Ferrero ranch was in the hands of his heirs. He died in February 1923 at the age of 67. The original walnut packinghouse, and now Puente citrus packinghouse, can be seen at the top center of the photograph. In the span of over a decade after his death, the ranch experienced some interesting incidents. A permitted and controlled fire got so out of hand on the ranch that the fire officials had to recruit men "from the Puente streets to serve as firefighters," reported the *Pomona Progress Bulletin* on November 8, 1926. Another report by the same publication on April 17, 1948, describes the account of Joseph Vincent Ferrero's witnessing of a minor crash landing of an airplane in one of the fields that he was plowing. The pilot was Russell H. Knott, of the Knott's Berry Farm family. He was on final approach to the Skyline Ranch Airport when he ran out of gas. Knott and his wife, Mildred, escaped unhurt. (Courtesy of the Perez family.)

As the walnut and citrus industries grew through the 1920s, Walter Temple had more trees planted on the Workman Homestead, seen here looking to the northeast around 1940. By the time this photograph was taken, the coddling moth may have done damage to some of the trees. Beyond the homestead, a vehicle is traveling north on Hacienda Boulevard, and the Joseph Ferrero ranch can be seen farther out. The growing town of Puente can be seen in the distance, and the Little Puente Hills beyond that. The long, sweeping line at the top right is the Southern Pacific Railroad and the pepper-tree-lined Valley Boulevard stretching east to Walnut, Pomona, and beyond. (WTFHM.)

By the late 1920s, Walter Temple was beset by diminished oil production from his wells and failed real estate projects. In 1930, Temple leased 20 acres of the Workman Homestead to Lawrence V. Lewis for his Golden State Military Academy for a term of 40 years. The academy offered students from the second to twelfth grades schooling in academics, military, commercial, and aeronautic fields. By 1932, Temple was unable to keep pace with mortgage payments and California Bank foreclosed on the property, keeping the lease to Lewis active for the school. The academy operated at the ranch for three more years, offering more courses as time went on, and changed its name to Raenford Military Academy in 1933. Two years later, it moved to the shuttered Encino Country Club. This photograph portrays an assembly of students in front of the Homestead's water tower in 1932. (WTFHM.)

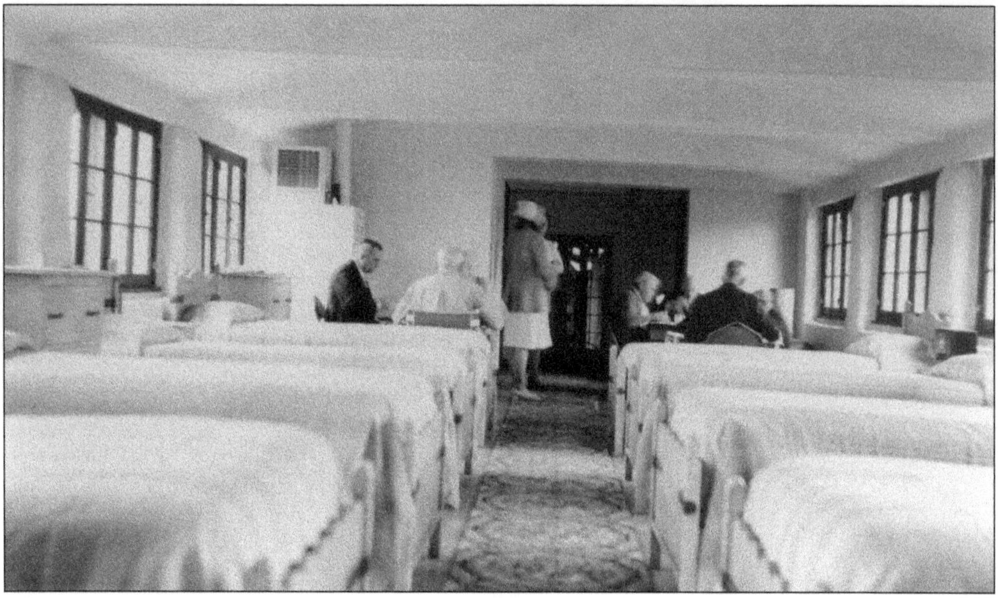

From 1935 to 1940, California Bank had caretakers at the Homestead who were from the Brink family, as in the Brink's security company. Harry and Lois Brown bought the property in 1940 and moved their sanitarium for the mentally and physically infirm from Monrovia and named it "El Encanto." The Workman House was at first used for the living quarters of the Browns, including their three sons, and eventually, it became the facility's offices and nurses' station. La Casa Nueva included patient quarters and recreation and entertainment rooms, while the old winery buildings turned auditorium and cafeteria were used for those same purposes. Pictured above is one of La Casa Nueva's wings used as a dormitory. Seen at the reservoir/swimming pool are two unidentified nurses and Ken and Gene Brown, sons of the owners. These photographs were both taken in the 1940s. (Both, WTFHM.)

In the 1940s, the La Puente Valley was still a rural region. With most of the industrial interests still located to the west, the farms, ranches, and small towns of the area endured. Seen here on what is today named Colima Road and looking east from about Azusa Avenue is the 1938 Model A Flathead Ford of a teenaged Buddy Parriott, around 1942. To the left of the road is where the Puente Hills Mall now stands, and on the right is today's Supervisor Pete Schabarum Park and more retail stores. Formerly known as Fifth Avenue, it was a regionally planned route in the 1920s that was built to relieve the congestion that existed on Valley Boulevard, the main road between Los Angeles and Pomona. (Courtesy of the Parriott family.)

August Von Handorf immigrated from Germany in the early 1880s, and by 1894, he had started his first dairy in Highland Park. After moving the operation to a few different locations within the cities of Los Angeles and Burbank, Handorf bought almost 700 acres in what would become the City of Industry and Hacienda Heights for his dairy. Bernard Von Handorf was born in 1899 to August and his first wife, Rose Kopp. Known as Ben, he and his wife, Mary, lived for over 40 years in a house on Turnbull Canyon Road, which still stands today, and he is the namesake for B.V. Handorf Drive. This c. 1940 photograph shows the dairy that was located near the intersection of Clark and Seventh Avenues, looking northeast toward the Little Puente Hills. In 1957, about 300 acres of the dairy property were sold to the Union Pacific Railroad for industrial development. (COI.)

Two distinct features of this late 1940s aerial photograph are the elongated oval of the Sky Ranch Airport in what is now the city of La Puente, at the top left, and another oval at the bottom center, which was a horse racetrack that never hosted a sanctioned event. The public airport existed from the mid-1940s until the early 1950s, when the wane of the private flying industry and rising land values combined to motivate the property's owners to develop housing. The racetrack was graded in 1943 by the Southern California Jockey Club and its president, Al Uniak. Approval by the state racing commission was never granted, despite a long legal battle that lasted until 1950. Today, it is the location of the Puente Hills Mall and State Route 60, Pomona Freeway. (WTFHM.)

By the late 1940s, the walnut production in the La Puente Valley was declining but still viable, as seen here in 1948 on a fully loaded REO Speed Wagon, owned and operated by Samuel J. Parriott, at the La Puente Valley Growers Association packinghouse. The packinghouse was probably close to its end of operations at the time of this photograph, as what remained of later harvests was processed in San Fernando and Chino. As seen here around 1949 with mechanic A.J. Lewis and an unidentified observer, Parriott's spare time was spent competing in motorcycle hillclimbing, flat track, and speed record events across the western states. He would later compete on the National Hot Rod Association circuit, driving the City of Industry special racecar to championship wins and speed records. (Both, courtesy of the Parriott family.)

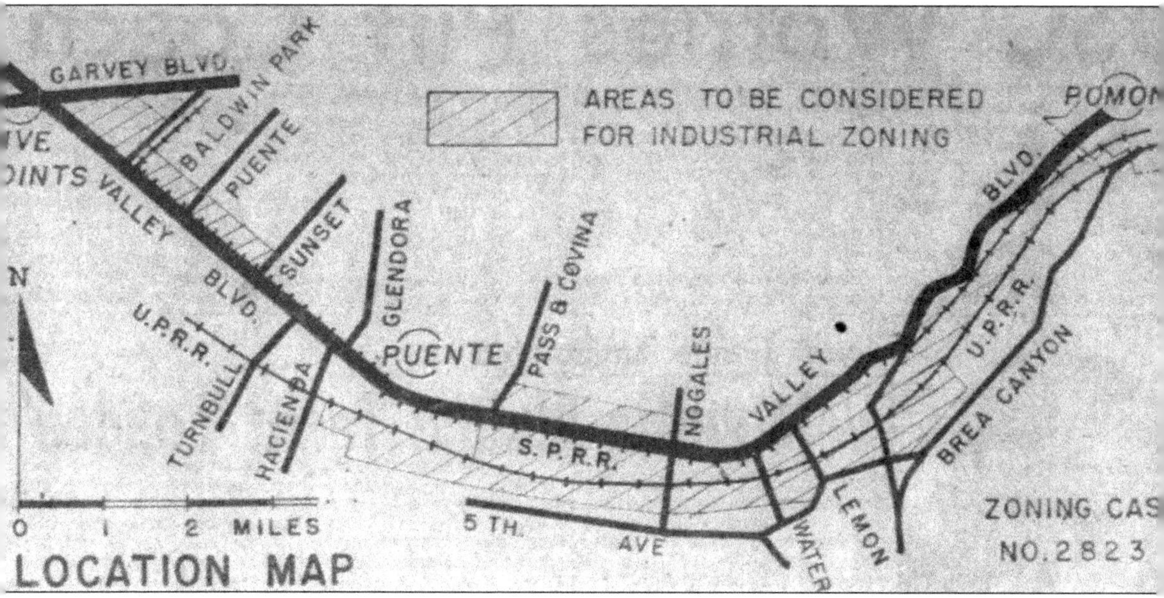

When the County of Los Angeles saw that widespread development of the East San Gabriel Valley was imminent, it tasked the Regional Planning Commission with producing a comprehensive plan of development for the area. Commission director Milton Brievogel estimated that a million people would ultimately reside in the East San Gabriel and Pomona Valleys, and he used that figure to calculate the ratio of one acre of industrial land for every 116 residents in order to provide enough jobs for the expected population increase. The September 1954 map represents the 3,500 acres of mostly uninhabited land that the commission proposed to add to the already existing industrial zoning of the valley. The dominant zoning at this time was designated for agriculture and the proposed zone was M1 ½, limited heavy industrial use, while smog-producing industries, such as smelters and foundries, would be prohibited. Brievogel cited the location between the two major railroads and protection from residential zoning as the main reasons for placement of the industrial zone. (*Pomona Progress Bulletin*, September 22, 1954.)

With the county's designation of the new industrial zone complete, local communities eyed the land for inclusion to their spheres of influence. In July 1955, members of the La Puente area began incorporation proceedings, which included the county's newly zoned industrial land as well as their residential base. The owners of the industrial-zoned properties protested inclusion to the proposed City of La Puente, preferring to avoid the possibility of a residential municipality dictating how their properties might be rezoned, taxed, and conflict with a bedroom community. The board of supervisors deleted those properties, and the remaining land was successfully incorporated as the City of La Puente in August 1956. This 1947 H&H photograph of La Puente's Main Street is looking east from its intersection with Glendora Avenue. (WTFHM.)

Two

Incorporation and Expansion

Graphik Circuits was a manufacturer of electronic circuit boards for the Minuteman, Sergeant, and Titan Missile programs, as well as the Mercury, Gemini, Apollo, and Mariner space projects. Graphik moved to the area named Electronics Park on Turnbull Canyon Road in February 1957. Engineering, research, and manufacturing operated out of this facility. Production continued at the location until 1989, when it was known by the name of TRW Benchmark. (COI.)

CITY OF LA PUENTE
J. T. BASSETT'S SUBDIVISION

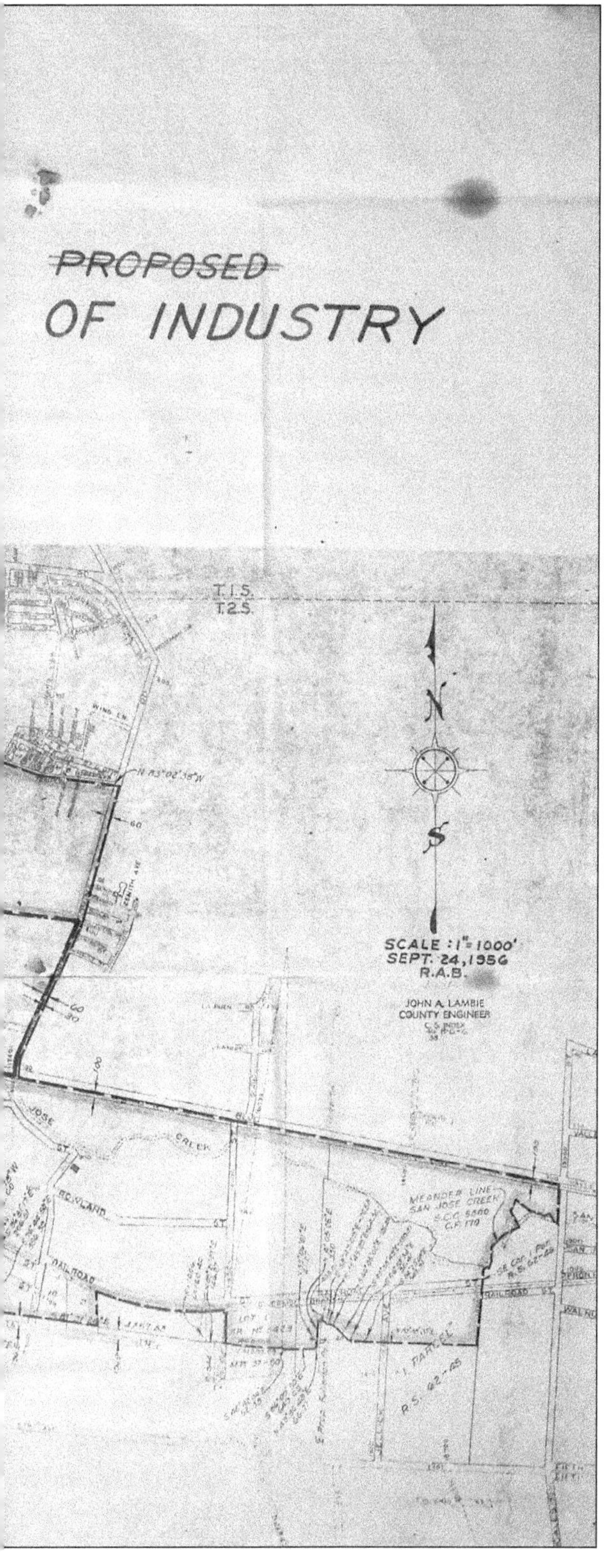

Soon after the La Puente incorporation was complete, the landowners who protested their inclusion to that city began the process of forming their own municipality. Among the largest landowners on the incorporation petition were the Southern Pacific Railroad, Sunkist Corporation, Clara Baldwin Stocker, James M. Stafford, and Edward Lustgarten. Co-owner of the El Encanto Sanitarium, Gene Brown, was quoted as saying, "The feeling is that we have a better chance for survival as a purely industrial section if we keep the area in one piece. Otherwise, zoning would be jeopardized, and parcels subdivided for housing developments." Local farmers and residents supported the incorporation and an overwhelming majority voted to incorporate on December 4, 1956. This image is the original Proposed City of Industry Boundary Map from September 24, 1956. (COI.)

Soon after the election, there was a lawsuit that challenged the minimum number of 500 people that was required to live within the boundaries to qualify for incorporation. The Superior Court ruled against the plaintiff, and an appeal was eventually dismissed, allowing for the incorporation to be certified on June 18, 1957. The organizational meeting of the city council was held two days later in a conference room at the Utility Trailer Company. Among the initial actions taken were electing John Ferrero as mayor and Samuel L. Parriott as mayor pro tem, adopting the city manager form of government, and executing a general services contract with the county. The room was reportedly overflowing with visitors including well-wishers from the cities of La Puente and Baldwin Park, county administrators, and residents. This photograph was taken at that meeting and shows county clerk Gordon Nesvig administering the Oath of Office to, from left to right, Darius Johnson, John Ferrero, Vera Zabaldano, Charles Rowland, and William Tallon. (*Daily Tribune*, June 21, 1957.)

Shown here around 1960, John Ferrero was not of the Joseph Ferrero family but was also born to Italian immigrants, Pietro and Theresa, in 1912. The couple moved around the area, following the jobs and having children along the way. In 1921, they found 30 acres in what is now the City of Industry. When Pietro died at only 48 years old, John and his mother were left to work their alfalfa fields and walnut orchard. Later, John farmed the Old Ranch area of Brea Canyon Road near the city of Walnut and, in 1962, leased the Diamond Bar Golf Course from the county to pasture his cattle. Even after surviving an automobile collision with a train in 1985, "Mayor John" continued his farming and civic service until his death in 1996. He was elected the first Mayor of the City of Industry and remained so for 39 years, using a no-nonsense approach to the job of turning the City of Industry from farmland into a regional economic center. (COI.)

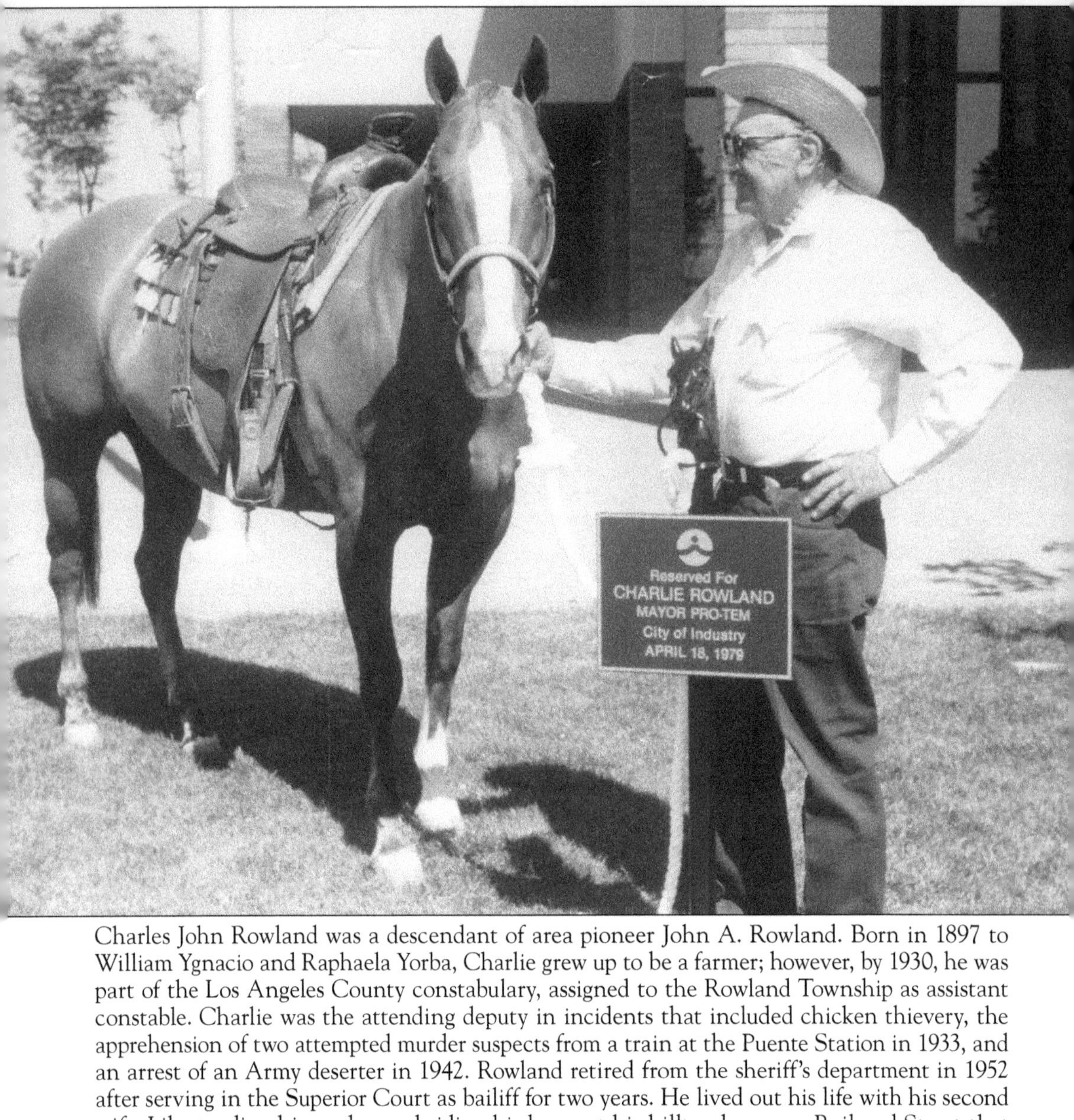

Charles John Rowland was a descendant of area pioneer John A. Rowland. Born in 1897 to William Ygnacio and Raphaela Yorba, Charlie grew up to be a farmer; however, by 1930, he was part of the Los Angeles County constabulary, assigned to the Rowland Township as assistant constable. Charlie was the attending deputy in incidents that included chicken thievery, the apprehension of two attempted murder suspects from a train at the Puente Station in 1933, and an arrest of an Army deserter in 1942. Rowland retired from the sheriff's department in 1952 after serving in the Superior Court as bailiff for two years. He lived out his life with his second wife, Lila, tending his garden and riding his horse at his hilltop home on Railroad Street that looked out over a greatly altered section of his family's Rancho La Puente. Charlie is shown here in 1979 at the Industry Hills Conference Center next to the hitching post installed in his honor and the ranching life that he represented. (COI.)

Robert Rope was a civically active person during his early life, from being a Boy Scout to being selected Long Beach Junior College Outstanding Man. Bob went on to earn degrees in history, education and public administration from the Universities of California at Berkley, Los Angeles, and Southern California. He was a teacher at Chaffey Union High School from 1939 to 1943 and was a lieutenant in the US Navy during World War II. In 1948, Rope was hired by the City of Long Beach as an investigator for the prosecutors' office and was engaged in several civic organizations including the Long Beach Sports Boosters, American Legion, Southwest Aviation Commandery of the Naval Order and the City Armed Services Commission. When hired by the City of Industry in 1957, Rope was an executive assistant at the County Regional Planning Commission. He guided early city councils through the process of establishing the municipality as the dynamic economic force that it soon became. Bob served the city for 14 years and passed away in 1975. (COI.)

Women have occupied important roles in the City of Industry from the very beginning. Included on the ballot for Industry's incorporation and its city council were the city's treasurer and clerk positions. Filling those roles and shown here in 1958 are Vera Zabaldano, treasurer (left), and Elvira (Vida) Ward, clerk. Ward was a member of the pioneer Rowland family and married to Joseph A. Ward. She held the office of the city clerk until shortly before her death in 1972. Zabaldano was the granddaughter of Joseph Ferrero, from Puente, and worked as a grader in the citrus packinghouse as a young woman. She lived in the City of Industry with her husband Joseph B. Zabaldano, a captain with the Los Angeles County Fire Department. In 2002, Rolene Harrison was appointed as the city's first female council member after serving on the planning commission and redevelopment agency's board of directors. In the 2019 city council reorganization, fellow council members selected Cory Moss to become the first woman to occupy the mayor's seat. (Courtesy of the Perez family.)

Glenn R. Watson, along with his law partner Robert Beverly, played a major role in the municipal incorporation boom of the 1950s, serving as city attorney for many of the new municipalities in Los Angeles County. He is pictured here in 1957 with, from left to right, John Ferrero, Watson, Darius Johnson, Charles Rowland, Vicente Perez, Joseph Alzugaray, Robert Rope, William Tallon. (COI.)

Robert Beverly was Industry's first assistant city attorney, while also serving as a Manhattan Beach City council member. He was elected to the state assembly in 1968 and coauthored the Song-Beverly Consumer Warranty Act in 1970, which paved the way for laws such as the "Lemon Law" protecting buyers of faulty automobiles. He is shown here being sworn-in at the City of Industry council meeting of April 15, 1958. (COI.)

The photograph above shows city council members in 1957 wearing honorary badges that the Los Angeles County Sheriff's Department issued to all city's council members. From left to right are Frank Bletcher (sheriff's inspector), Darius Johnson (council member), Joseph Alzugaray (council member), Eugene Biscaluiz (Los Angeles County sheriff), Les Haight (sheriff's inspector), Mayor John Ferrero, Charles Rowland (mayor pro tem and former Los Angeles County deputy sheriff) and William Tallon (council member). This was the second version of the council, as original council member Samuel L. (Buddy) Parriott resigned in August as a result of his change in residency. The photograph below is of the council and staff seated at the dais in the conference room of Utility Trailer Company on April 15, 1958, with newly elected council member Filbert Rowland (in white shirt). (Above, COI; below, courtesy of the Perez family.)

These two views show the house that belonged to Teen Ferrero and was used for the core of the city's first city hall, and the completed building, both photographed in 1958. The location was at the southwest corner of Hacienda Boulevard and the Southern Pacific Railroad and would remain there until 1975. Parts of the Stafford Mill and the walnut packinghouse can be seen to the left of the house on the north side of the railroad tracks. (Both, courtesy of the Perez family.)

Among the many items that the early city councils considered were numerous annexations of properties into the new municipality. By October 1957, there were almost 800 acres of annexations being considered or in the process of certification. The city's first approved annexation was about 10 acres on Baldwin Park Boulevard, north of Torch Street (Amar Road). In 1957, the city's area was a little over five and a half square miles, and its 53 businesses had a combined payroll of $18.9 million. By 1963, the city grew to nine and three-quarter square miles, and the payroll leapt to over $50 million. The participants in this 1958 council meeting inside the new council chambers are, from left to right, Elvira Ward, Bob Rope, Darius Johnson, John Ferrero, Glenn R. Watson, Filbert Rowland, and Charlie Rowland. (Courtesy of the Los Angeles County Fire Department.)

In 1955, the Carrier Corporation bought 68 acres on Anaheim-Puente Road from the Fred H. Bixby Ranch Company and constructed a 500,000 square foot, $6.5 million manufacturing facility to produce heating, air-conditioning and water heater units under the Payne and Day & Night divisions. The plant employed 980 people and required two miles of conveyor belts. The company and its employees were active in Red Cross Blood Bank drives, and its executives were actively involved in the chamber of commerce. Carrier still operates a Transicold service center on the property. (Both, COI.)

While many industrial zones of the greater Los Angeles area were utilitarian in appearance, it was important to the City of Industry that businesses establish and maintain attractive grounds at each site. The Eagle Manufacturing Company, shown here around 1962, was recognized by the Los Angeles Beautiful organization in 1964 for its building and landscaping. The manufacturer of light fixtures moved from Los Angeles to a 10-acre site on Turnbull Canyon Road in 1961. Other local companies receiving recognition that year were the Gaylord Division of Crown Zellerbach and Fruit Growers Supply, both located on Baldwin Park Boulevard. The city and chamber of commerce would, in later years, honor companies that enhanced the appearance of their property beyond the city's own requirements. (Both, COI.)

Lyon Metal Products was founded in 1901 in Aurora, Illinois, and opened a facility in the City of Industry in the late 1950s. During World War II, Lyon manufactured products that included bombsight cases, tail assemblies for fighter aircraft, and tank parts. The company and the entire nation experienced a sheet metal shortage after the war's manufacturing boom, and one of the ways that Lyon dealt with the situation was to offer potential customers finished products equal to the weight of a customer's scrap or used metal. Its Plant No. 3 was located on Turnbull Canyon Road south of the Union Pacific Railroad, shown here around 1960. The company manufactured all types of metal shelving, lockers, shop, and kitchen equipment from this location. Lyon was awarded many local contracts for outfitting educational facilities with locker units; however, its operations ceased at this location about 1976. (Courtesy of the Perez family.)

In 1962, the Owens-Illinois Glass Company opened the Libbey Glass tableware manufacturing plant on Anaheim-Spadra Road (now Brea Canyon Road). This site was chosen from more than 150 other locations because of what company executives said were overall better labor conditions and attractive gas, fuel, and transportation facilities. The factory employed more than 200 people with the capacity to produce 100 million glasses annually. In the above photograph, the facility is juxtaposed against a small herd of grazing cattle in a neighboring pasture, depicting the progression of old farmland into industrial usage in the 1960s. The photograph below was taken in 2002 at the plant's 40th anniversary with, on the left, Don Sachs (executive director of the Industry Manufacturers Council) congratulating Willie Purvis (Libbey Glass). At that point, the plant employed 250 people, producing 250,000 tableware pieces per day. (Both, COI.)

In the Little Puente Hills that is bordered on the east by Pass and Covina Road (Azusa Avenue), there were two refuse disposal sites that operated amid the hill's slopes and canyons. In 1956, Puente Waste Disposal Company's location was shown on a rudimentary map and advertised as the "New Dump" in the area. In December 1957, the city council of the newly incorporated City of Industry approved a permit for the other company, Refuse Disposal Corporation, to operate a facility next to the existing landfill operations. These two photographs, taken by the Los Angeles County Fire Department in 1961, examined the problem of refuse fires that had occurred frequently at the two operations. In 1963, the city had made the decision to study part of the hills for recreational use. (Both, COI.)

The City of Industry was established as a contract city, meaning it utilizes county resources for policing, fire protection, building and safety, public works, and other necessary services. This form of municipal operation is often referred to as the Lakewood Plan. In April 1954, the newly incorporated Los Angeles County City of Lakewood adopted the contracted services type of government that allowed the city to begin operating with minimal financial outlay. The county had been providing some of its essential services to incorporated areas since the early 1900s; however, Lakewood was the first to employ them in an extensive manner. This was largely a response to the enormous population growth after World War II, and the desire to have local administrative control along with the necessary services that a municipality requires. Shown here at the 1962 groundbreaking ceremony for the Industry Sheriff's Station are, from left to right, Joseph Licata (sheriff's captain), David Sprong (La Puente mayor), John Ferrero (Industry mayor), and Edwin Estudillo (La Puente–Industry Chamber of Commerce president). (COI.)

County supervisor Frank Bonelli participated in many events held throughout the city during its enormous early growth period and liked to refer to the City of Industry as the "Breadbasket of the San Gabriel Valley." Bonelli came from a business background, which may have endeared him to the city's mission of job creation. He is shown in the photograph above in February 1963 while touring the construction sites of Los Angeles County Fire Station No. 43 and the Industry Sheriff's Station. From left to right are Joe Dooley (La Puente–Industry Chamber of Commerce), Leonard Westhoff (Walnut mayor), Frank Bletcher (sheriff's inspector), John Ferrero (Industry mayor), Frank Bonelli, Keith Klinger (fire captain), Robert Henger (La Puente–Industry Chamber of Commerce), James Downey (undersheriff), and David Sprong (La Puente mayor). In the photograph below, Bonelli is joined by Mayor John Ferrero and fire captain Keith Klinger. (Both, COI.)

In the 1960s, the City of Industry saw many county support services increase their presence within its boundaries. With the increase in the number of businesses locating to the city, as well as residential numbers in the surrounding communities, the firefighting and policing agencies had to keep pace. The 1963 grand opening of the Los Angeles County Fire Station No. 43 on Stimson Avenue is pictured above. The attendees are, from left to right, Samuel J. Parriott (council member), Frank Bonelli (Los Angeles County supervisor), John Ferrero (mayor), Charles J. Rowland (mayor pro tem), Keith Klinger (fire captain), Filbert Rowland (council member), and two unidentified representatives from the nearby Veterans of Foreign War Post No. 1944. The photograph below shows the station and vehicles in 1968. (Both, COI.)

As health standards changed in the State of California, the Brown family were unable to continue operating El Encanto out of the historic Homestead structures. The Browns sold off portions of the Homestead's 92 acres in order to finance the new El Encanto facilities adjacently north of the Workman Homestead. This photograph illustrates the 1963 groundbreaking ceremony and includes, from left to right, Lesley Dagley (El Monte mayor), David Sprong (La Puente mayor), Robert Brown (El Encanto co-owner), Leonhard Westhoff (Walnut mayor), and John Ferrero (Industry mayor). This was the first of a multiphased project of modernization and expansion for the facility and consisted of 80 beds in private and semiprivate rooms in a total area of 25,000 square feet and constructed at a cost of about $500,000. The second phase that added 168 beds and all modern furnishings and equipment was completed in 1969, and the owners donated 169 pieces of the older furnishings to local families in need. (COI.)

The Los Angeles County Sheriff's Department served the City of Industry from the San Dimas Station during the early part of the city's existence. As the industrial and residential population was booming at that time, there was greater need for a local base of policing. The city purchased just over three acres of property owned by the C.C. Stafford Mill Company in late 1961, for the express purpose of donating the land to the county for a new sheriff's substation. An extension of Abbey (Stafford) Street was also constructed and paid for by the city for access to Hacienda Boulevard. In February 1964, the 24,000 square foot, $625,000 facility was opened and staffed by 125 department personnel headed by Capt. Joseph Licata. Seen here at the February 27, 1964, dedication ceremony are, from left to right, Licata, David Sprong (mayor of the City of La Puente), Frank Bonelli (Los Angeles County supervisor), unidentified (mariachi), James Downey (undersheriff), John Ferrero (mayor of the City of Industry), and Leonard Westhoff (mayor of the City of Walnut). (COI.)

In the 1930s, the idea of building freeways was used to create jobs, while making a more efficient way to get around the region. A comprehensive plan for a county highway system was developed in the 1950s to relieve automobile congestion on local streets. Included in these plans was the San Gabriel River Freeway that was a route from the San Gabriel Mountains in the north to the Long Beach area in the south. The designated Interstate 605 was adopted in 1954 and estimated to cost $60 million. In 1964, a four-mile section of the freeway between Valley Boulevard and Whittier Boulevard was completed and ceremoniously opened by, from left to right, Fred Bagshaw (state public works department), Chuck Wiggins (El Monte mayor), Ruth Benell (Pico Rivera mayor), John Ferrero (Industry mayor), David Sprong (La Puente mayor), and Edward Telford (division of highways). The ribbon-cutting ceremony was performed with an acetylene torch and ribbon that were both produced by companies in the City of Industry. (COI.)

Hacienda Boulevard is seen from above looking north in this c. 1964 photograph, taken shortly after its widening and separation from the grade crossing of the Union Pacific Railroad. The construction was part of a county master plan highway project. Starting at the bottom is a residential area of Hacienda Heights, just to the south of the railroad running left and right, while the San Jose Creek, still in its natural state, flows to the west toward the San Gabriel River. Hacienda's intersection with Valley Boulevard is near the top of the photograph with only three corners of the intersection developed at that time. On the north side of Valley and east of Hacienda Boulevards at Russell Street is a miniature golf and amusement park. Farther north on Hacienda Boulevard, one can view the City of Industry's first city hall on the right, while the buildings of the old La Puente Valley Walnut Association's Packinghouse and the C.C. Stafford Mill are on the left. (Courtesy of the Perez family.)

This aerial view, taken about 1964 and looking north, reveals the railroad infrastructure that was in place before many of the industries who would use them were built. The curving lines of the railroad's spur tracks course through this area near Ninth and Proctor Avenues, which is in the upper-right portion of the photograph. Near that intersection, one can see a natural gas facility, which no longer exists, and a few industrial developments. At the bottom is the former San Pedro, Los Angeles & Salt Lake Railroad, now the Union Pacific, and Salt Lake Avenue just above the tracks. San Jose Creek can be seen laterally through the center, and making its way down to empty into the San Jose, Puente Creek is at the top left. (Courtesy of the Perez family.)

Flooding in Los Angeles County has always been an issue during extreme rainy periods. After flowing down from heights of up to 10,000 feet elevation in the San Gabriel Mountains, rainfall still flows rapidly through the basin, dropping at a rate of 25 feet per mile. After a particularly damaging flood event in 1914, the county passed the flood control act and formed the county flood control district just one year later. In 1917, a bond measure passed for $4.45 million to begin construction primarily on dams in the mountain canyons. Another issuance of bonds for $35.3 million in 1924 further financed the districts continued construction of dam structures and channel lining. In a 1930 report to the district, a total of 523 miles of waterways were deemed to be under the county's purview. One of those waterways is the San Jose Creek, shown here around 1964. Note the flood control district's lining of the creek with cement near the intersection of Sixth and Lomitas Avenues. (Courtesy of the Perez family.)

In 1961, the city council authorized city attorney Glenn Watson to negotiate the acquisition of the Workman Home and El Campo Santo Cemetery for historical preservation. The Workman Home and La Casa Nueva were still owned and being used as El Encanto Sanitarium by the Brown family, who were planning a new location just north of the historic site. The Browns were sensitive to the historic significance of the buildings since their purchase of the property in 1940. In December 1963, the city executed an agreement with the Browns in which they were given five years to continue their operations and yet provide personnel for scheduled public tours and inspections of the site at no cost to the city. The northeast corner of the Workman Home is shown here in 1966 with Temple descendant Edith Temple Stanton. (WTFHM.)

The Walter P. Temple Memorial Mausoleum is located inside El Campo Santo cemetery and is shown here in 1966. This burial site holds the remains of individuals that had profound influences on the formation of the greater Los Angeles area and the state of California. Along with the Workman Homestead founders, William and Nicolasa Workman, there lies the reinterred remains of the last governor of Mexican Alta California, Don Pio Pico, and his wife, Maria Ignacia Alvarado. The mausoleum was built by Walter Temple after he purchased the homestead in 1917. The first recorded burial in the cemetery was that of Workman's older brother, David, in 1855. Along with Workman's co-owner of Rancho La Puente, John Rowland, there are several others buried there that are from local pioneer families. (WTFHM.)

During the second ever City of Industry City Council meeting, June 27, 1957, council member William Tallon complained of a 30-minute delay experienced by attendees of the meeting at the railroad crossing of Anaheim-Puente Road and the Southern Pacific Railroad. A decade later and after several demonstrations of promise and retraction from the county and state, a joint effort between those entities, Southern Pacific, and the cities of La Puente and Industry resulted in a completed six-lane overpass. It was another county master plan highway improvement that benefitted travelers between Orange County and the San Gabriel Mountains. The 740-foot steel girder, concrete deck structure spans the widths of Chestnut Street, San Jose Creek Flood Control Channel, Southern Pacific tracks, and Valley Boulevard. It came at a cost of $2.25 million, shared by the above-mentioned stakeholders. Pictured in April 1967 are Industry council members Samuel J. Parriott and Darius Johnson observing the final phases of the construction from the intersection of Chestnut Street and Virgil Waters Way. (COI.)

The first section of the Pomona Freeway was constructed in East Los Angeles in October 1965. The portion that traversed the western part of the City of Industry from the San Gabriel River Freeway (Interstate 605) to Fullerton Road was dedicated on October 19, 1967. The above aerial view shows the intersection of the two freeways looking west toward the San Gabriel River in the late 1960s, and the view below is of the Pomona Freeway at Hacienda Boulevard. The dedication ceremony for this segment of the freeway was sponsored by the City of Industry and took place in the eastbound lanes at the Workman Mill Road bridge. Construction was completed in about 21 months by Peter Kiewit Sons' Company, at a cost of $8.6 million. (Both, COI.)

The next section of the Pomona Freeway was completed and dedicated in May 1970. The six-mile, $10 million segment runs from Fullerton Road in the City of Industry to Diamond Bar Boulevard in the City of Diamond Bar. A crew from the E.L. Yeager Construction Company is shown in these two photographs, around 1969, building the support system for one of the five undercrossing structures for that portion of the freeway. Two more segments were slated to be completed by the end of the summer that would extend the route to the City of Ontario. (Both, COI.)

In 1967, new construction in the city totaled $10 million, the highest annual figure achieved to that point. The city gained 18 new businesses in the year of 1968, bringing the total number to 255, up from about 50 when the city was incorporated 10 full years prior. The city now covered 10 square miles, and only about 20 percent of it was developed. The number of people employed in the city was 24,000, and the total payroll was $180 million. Ralee Engineering was one of those 18 companies that opened that year with its location on Gale Avenue in the Bixby Industrial Park. Ralee was a manufacturer of aircraft parts for an industry that had 247,000 employees in the Los Angeles and Long Beach areas. That was the highest percentage (17.7) in the nation, in a $27.2 billion aerospace industry in 1967. (COI.)

Mattel, Inc., was considered the world's largest toy manufacturer when it opened its new $1.5 million, 420,000-square-foot facility in the City of Industry in January 1964. The plant was located on 19 acres at Temple Avenue near Baldwin Park Boulevard and employed 1,000 people in 1964 with plans to increase that number to 1,700. Mattel founders, Elliot and Lois Handler were honored at a La Puente–Industry Chamber luncheon in July 1964, where Elliot Handler lauded the city for its business climate and the San Gabriel Valley for its industrious labor force, which allowed Mattel to commence operations with fewer problems than he had anticipated. By 1967, Mattel expanded the facility to 670,000 square feet and totaled 2,100 employees, and in that same year, Ruth Handler was named president of the company. This expansion made it the largest of the company's plants. Shown here are some of the employees exiting the facility in 1968. Mattel added a 500,000-square-foot distribution center at Valley Boulevard and Stimson Avenue in 1969. (COI.)

During the time of the city's transition from open farmland into attractive and usable industrial properties, it was necessary to address unkempt and, at times, dangerous property conditions. This photograph illustrates such conditions on acreage at Nogales Street, south of the Southern Pacific Railroad in 1969. The situation was brought before the City of Industry Rehabilitation Board, which acted on the recommendation of Los Angeles County Building and Safety and fire department to call for the removal of all debris. This land was part of the largest annexation of property by the city in 1957. A total of 549 acres was added to the city's girth after a petition of annexation was submitted to the city council by landowners, including Roy Benton of Benton Feed Yard, in what was then considered the area of Walnut. (COI.)

Three

Sustained Growth and Diversification

Inside the new city hall council chambers and seated at the dais, c. 1975, are, from left to right, Betty Lynn (back to camera), Katherine "Kitty" Phillips (deputy city clerk), Robert Rope (city manager), Darius Johnson (council member), John Ferrero (mayor), John Paul Ferrero (council member), Martin Becker (city attorney), Samuel J. Parriott (council member), and Charles Rowland (mayor pro tem). (COI.)

Even though the decade of the 1960s was booming with new industrial buildings in the City of Industry, there were still large tracts of undeveloped land available. By the end of the decade, the total assessed valuation of the city was $125 million, up from $37 million in 1961. The above photograph shows a vacant area at the intersection of Don Julian Road and Ninth Avenue, looking southwest toward the Puente Hills, and the photograph below is of the Bixby Industrial Park site, looking northeast from Gale Avenue. Both photographs were taken in the late 1960s. (Both, COI.)

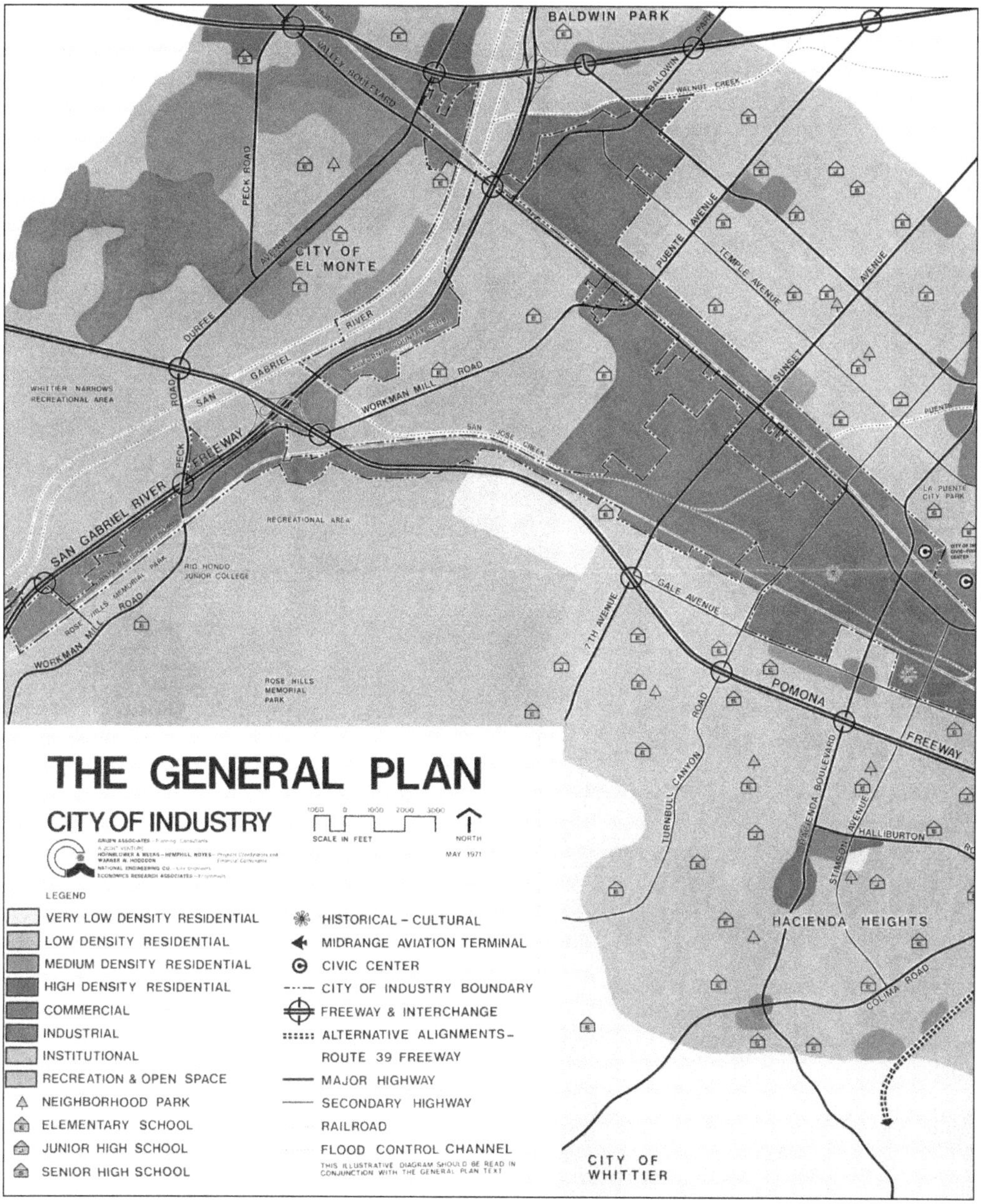

The first goal that was stated in the city's original general plan was the following: "The maintaining and further development of an employment base in the San Gabriel Valley and Los Angeles metropolitan area." A general plan is intended to assist a city and its citizenry in formulating and adopting the city's aims of orderly growth, sound fiscal policies, and enhancement of existing values. This image is a portion of the map that accompanied the plan that was conducted for the city by Gruen Associates in conjunction with National Engineering and others and completed in 1971. The plan outlined the many unique characteristics and objectives that the city possessed and recognized the need for equally unique approaches to realize those goals. As the city did, in its infancy, consult the Stanford Research Institute's economic study, so too would it rely on the general plan to help guide its physical development to provide the much-needed jobs for the valley. (COI.)

The city's leaders knew that the best way to fulfill the goals of the general plan was to establish a redevelopment agency. The Industry Urban Development Agency was formed in April 1971, and a five-member board of directors was selected from the local community. The directors shown here in June 1972 are, from left to right, Raymond Reyes, Scella Harrison, Marynelle Bussola (chairperson), Frank Horne, and Manuel Garcia. The agency's executive director was Graham Ritchie, and the office was in the newly constructed redbrick building on Hacienda Boulevard. A city's redevelopment organization was allowed to issue tax allocation bonds to pay for acquisitions of properties; develop streets, sewers and storm drains; and sell the improved properties. The bonds were paid off from the increased property tax revenue on the improved land. The first project area was adopted in July 1971, and its first bond sale of $10 million was made in December to facilitate traffic circulation and address local drainage issues that prohibited building within parts of the project area. (COI.)

One of the reasons cited in the general plan as inhibiting development was the poor secondary drainage of large lots in the central part of the city. Los Angeles County was the lead agency for the construction of the primary flood control system that focused on the San Jose Creek; however, there was still localized flooding that the city was responsible for mitigating. By way of the city's redevelopment agency, funds were available to construct underground storm drainage in order to allow construction on otherwise unusable land. At right is a page from the Gruen Associates general plan study, titled "Typical Obstacles and Problems," and below is a 1970s photograph that shows a crew pouring cement in the construction of one of the drainage channels. (Both, COI.)

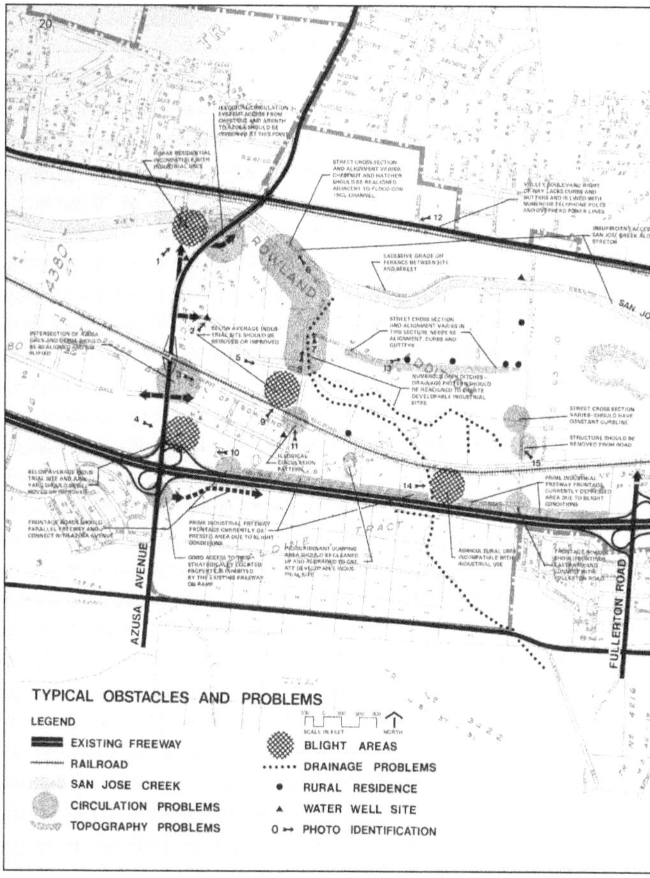

Facilitating the rapid growth of the city's industrial development was, and still is, the use of the tilt-up construction method. It was honed by Robert Aiken of Winthrop Harbor, Illinois, around 1908. His first structure was a target wall at the Camp Logan National Guard base and rifle range in Zion, Illinois. By pouring concrete wall panels on site and on the ground, then tilting them up when cured, it saved the time and expense of constructing them off-site or forming them in place. Aiken went on to construct housing, churches, and commercial buildings using this method. Shown here in the 1970s is the tilt-up construction of a typical industrial building in the City of Industry. (Both, COI.)

In 1961, a group of industrialists whose companies were in the City of Industry came together to organize a chamber of commerce that became known as the City of Industry League. The league sought to enhance the business-friendly environment where their companies thrived and continued to support the policies that the city had implemented. Among the businesses that pledged $500 each to the fledgling league were M and A Steel, McConnell Cabinet, Eagle Manufacturing, Hill Brothers, and Puente Ready Mix. The City of Industry League merged with the La Puente Valley Chamber of Commerce in 1963 and became the La Puente–Industry Chamber of Commerce. Los Angeles attorney Graham Ritchie was elected executive secretary of the league and would eventually become the city attorney and, for a while, its city manager. His tenure in these positions was during the seminal period when the city formed its general plan and redevelopment agency, where Ritchie was the first executive director. Ritchie was photographed here around 1969 at a chamber of commerce luncheon at the California Country Club. (COI.)

In 1968, the Industry Chamber of Commerce was created out of an organization called the Industry Manufacturers Association. Virgil Waters was elected as its first president, and Robert Henger was a director. Just one year later, the name changed to the Industry Manufacturers Council. Pictured in the c. 1970 photograph above is Waters, who was an executive at Utility Trailer Company as well as one of the early directors of the City of Industry League and a highly regarded civic leader. The city honored Waters with the naming of Virgil Waters Way, a street that is near Utility Trailer's Industry headquarters still to this day. Henger, left, was president of Eagle Manufacturing and was president of the chamber when it was partnered with that of La Puente's chamber. (Both, COI.)

Samuel Ichiyé (S.I.) Hayakawa was born in Vancouver, British Columbia, in 1906. Educated in both Canada and the United States, Hayakawa was known for his popular writings on semantics, and he taught English at different universities, including San Francisco State College (San Francisco State University). It was there that he was thrust into the limelight when he was tapped as acting president of the college after a period of student rioting in 1968. Hayakawa responded to the rioters with police force to restore order on the campus. He is photographed here at the California Country Club, addressing an Industry Chamber of Commerce luncheon in 1969, at which he expounded, with directness and humor, on his ideas of why the students took to violent measures against the education establishment. His views and actions found public favor as he remained president until 1973 and successfully ran for the US Senate three years later, serving one term. (COI.)

The Industry Manufacturers Council sponsored trips to Sacramento that allowed local businesspeople to meet with state lawmakers to address the needs of industrialists. In 1970, there were 13 members of the Industry business community and city staff that took part in the two-day trip. The primary topic reported to have been discussed were the increased tax burdens placed on manufacturers. The group met with Gov. Ronald Reagan, state director of finance Vernon Orr, and members of the California state legislature. Shown in the photograph above are, from left to right, Ronald Reagan (governor), Julian Lobosky (Metallurgical Testing Company), unidentified, Virgil Waters (Utility Trailer Company), and James E. Brown (California Steel and Tube). In the image below are, from left to right, Robert Henger (Eagle Manufacturing), Bill Campbell (California State Assemblyman), and Reagan. (Both, COI.)

Among the activities that the Industry Manufacturers Council sponsored was an annual bus tour of city developments and businesses. These photographs from 1970 show some of the 50 businessmen on the bus and touring the plant of Day and Night Manufacturing (below). From left to right are unidentified, Richard Lusk (Lusk Metals), Robert Henger (Eagle Manufacturing), F.H. Crouch (Sonoco Products), and unidentified. Included in this tour was seeing the progress of the Pomona Freeway construction at Azusa Avenue and a luncheon held at California Country Club with an economic outlook presentation from Franklin Stockbridge, a noted economist. (Both, COI.)

Through numerous property acquisitions made over more than a decade, the city completed its purchase of 500 acres in the Little Puente Hills in 1970 for the development of its long-sought park. And after 19 years of landfill operations on the property, there was an estimated three and a half million tons of solid and liquid waste buried. Problems such as ground subsidence from subsurface waste fires, poor drainage conditions, and trapped methane would have to be mitigated, and that process had to be monitored by four county and regional regulatory departments for the safe and proper development of the site. These c. 1970 photographs illustrate some of the landfill debris and soil erosion that existed during the engineering study phase of the Industry Hills project. (Both, COI.)

There were various uses that occupied the eastern area of the Little Puente Hills. In 1940, a zone variance was granted by the county to Dick Lautenbach for the development and operation of a dairy. Just one year later, there was a report of a hay fire at the dairy of Harold J. Grueter that required personnel and equipment from as far away as San Dimas to hold the destruction to 125 of the 500 tons of hay that were threatened. Oil was being produced at a rate of 150 barrels a day on land that was described in 1951 as the Lanford/Vanderhoof lease. Four years later, a housing tract with 250 lots on 59 acres gained county approval for L. Hansen and R. Vanderhoof, though this was not ultimately developed. The city approved a permit for an oil well in 1962 to Ramser Oil Company, and oil is still being pumped from a site on the property today. This c. 1968 photograph shows part of one of the dairies. (COI.)

In their natural state, the Little Puente Hills were a very good location for a recreational site. The type of soil, natural topography and lack of seismic faults were among the favorable engineering and geological conditions that geologist F. Beach Leighton summarized in an initial report for development. The unfavorable conditions that existed due to the landfill operations, however, were quite lengthy and their mitigations more detailed. These two late-1960s aerial photographs of the hills were taken looking northwest with Azusa Avenue in the foreground. The scarred areas of two landfills can be seen at the left and right portions of the hills. The photograph below is of one of the landfill operations. (Both, COI.)

What began in 1963 with a study for creating a city park of 300 acres had, by the end of the decade, grown to a 625-acre regional recreation facility. With the completion in 1975 of comprehensive engineering studies and plans of implementation, construction of the Industry Hills Civic-Recreation-Conservation development began the following year. These two c. 1976 photographs exhibit the grading of five million cubic yards of earth. During this initial phase of construction, there was also the unearthing and relocation of 225,000 cubic yards of refuse from the former landfill areas to other locations on site to allow the building of the exhibit/conference center and hotel structure. (Both, WTFHM.)

On what was once a walnut orchard, hayfield, and defunct horse racetrack, the Puente Hills Mall regional shopping center was constructed. The property at the intersection of the Pomona Freeway and Azusa Avenue was owned by Western Harness Racing, Inc., when, in 1972, the groundbreaking ceremony took place. Ernest W. Hahn, Inc., developed the mall, which cost about $40 million to build, on 94 acres, and it opened in 1974. This c. 1975 aerial view is looking northwest and shows the 1.3 million-square-foot structure with its anchor stores of Sears, the Broadway, JC Penney, and J.W. Robinson's. The parking lot near JC Penney was used in the 1985 film *Back to the Future*. On the left edge of the photograph is the construction of Otterbein Park (now Schabarum Park). (COI.)

With the growth of the central portion of the city came the need for more fire protection. Just north of the recently opened Puente Hills Mall, on the northeast corner of Azusa and Gale Avenues, Los Angeles County Fire Station No. 118 was built. Shown in the photograph above in 1975 at the groundbreaking ceremony are, from left to right, council members Charles Rowland and Samuel J. Parriott, Peter F. Schabarum (county supervisor), Richard Houts (county fire chief), and Scella Harrison (Industry Urban Development Agency chairperson). The station was built with funds from the city's redevelopment agency at a cost of $254,000 and was 4,200 square feet. The below photograph shows the station after its completion in 1976. (Above, courtesy of the Los Angeles County Fire Department; below, COI.)

The 20th anniversary of the city's incorporation was celebrated on June 18, 1977, with a presentation at city hall. Mayor John Ferrero is at the podium with Msgr. Thomas Kiefer, and council members Charlie Rowland, Darius Johnson, and Samuel J. Parriott in the background. There was much to recognize at this milestone event, as the number of businesses located in the city reached the 1,000 mark with 50,000 employees and a total payroll approaching $1 billion. Aside from the astounding numbers, there were the ancillary activities that the city council initiated to satisfy the elements of the general plan and to make it a well-rounded industrial municipality. The restoration of the Workman Homestead was underway, and the construction of the Industry Hills complex was also taking place. Along with presentations made by dignitaries, the 220 guests were witness to the unveiling in the lobby of two bronze plaques, created by local artist Rosalie Maus Bailey of Mayor John Ferrero and city manager Bob Rope. (COI.)

The main recreational attraction at Industry Hills are the two championship golf courses. Both courses were original designs of William F. Bell, a noted golf course architect who along with his father designed numerous renowned courses. The Eisenhower and Zaharias courses have won several awards and have hosted top professional and amateur tournaments from the very first year that they opened. Sally Little won both the 1981 and 1982 Ladies Professional Golf Association's Olympia Gold Classic tournaments held on these courses. Occupying part of the conference center was the Ralph Miller Golf Library, one of only two public libraries in the United States that were dedicated to the game at that time. These c. 1982 photographs depict a scene during one of the ladies' tournaments on the Eisenhower course and an aerial view of the courses. (Above, COI; below, WTFHM.)

One of the unique features of the Industry Hills complex was the display of the railroad car that carried Sir Winston Churchill's body to its final resting place in January 1965. The idea to acquire "Bogie Luggage Van s2464" was council member Darius Johnson's, who, a year earlier, had taken a trip to England and was a friend of Dr. Horace King, deputy speaker of the British House of Commons. Just a few months after the coach performed its historic duty of carrying the former prime minister from Waterloo station to the station at Handborough, which saw multitudes of English residents honoring Churchill from rail platforms and farm fields, British Railways planned to scrap the car. Upon hearing these plans, Johnson lobbied the rest of the council and negotiated an offer of $980 to purchase the coach and create a memorial museum to Churchill at the planned park site. It is shown here in September 1979 being transferred from the Port of Los Angeles. (COI.)

Even though this piece of British history was purchased and delivered to the Port of Los Angeles by November 1965, Johnson did not live to see the final presentation of BLV s2464, as the development of Industry Hills took another 14 years. It is shown here, around 1979, at its display location next to St. Andrew's Station during site construction and with the finished installment. Had council member Johnson not persuaded the city to obtain the Churchill coach, it would have been lost to a scrapper's torch long ago. After more than 20 years on site, interest abroad increased for the historic rail piece, and s2464 found its way back home to Britain, intact and ready to receive a new audience of admirers. (Above, COI; below, WTFHM.)

Another of Industry Hills' original features was a golf cart funicular. Planned, designed, and manufactured by the Voest-Alpine company in Austria, each car carried carts and golfers 140 feet up the hillside from the course to a landing at St. Andrew's Station snack bar, modeled after the railway station in Scotland. At the time, it was reported to have been the only funicular railway of its kind in the Western Hemisphere. After about 20 years of service, the operation and maintenance costs were determined to be too burdensome; however, it still sits on its hillside perch. These two c. 1982 photographs show the funicular and its hilltop destination. (Both, WTFHM.)

In July 1980, the Industry Hills Swim and Tennis Center opened featuring an Olympic-sized pool, diving platforms, springboards, and 17 tennis courts. The swim center hosted international events and attracted world-class swimmers to its Industry Hills Swim Club. Two of the club's members were part of the 1984 US Olympic team, and the next year, the club was ranked No. 4 in the country. The tennis center attracted events such as the nationally televised Michelob Light Cup exhibition tournament in July 1982 that featured world No. 2 Jimmy Connors versus No. 4 Bjorn Borg in a five-set final won by Connors. Above is the 1986 US Swimming National Junior Olympic Championship, and below is the 1982 tennis tournament. (Both, COI.)

Even before the Industry Hills Sheraton Hotel was opened in January 1982, the 625-acre, $65 million project was named Outstanding Civil Engineering Achievement for 1981 by the American Society of Civil Engineering. The city's civil engineering firm of National Engineering Company was the project coordinator. This award recognized the innovative accomplishments of turning a waste disposal site into a world-class destination and host of premier sporting events, all while being open to the public. Among the innovations was the use of the former landfill's methane gas, extracted by wells and routed via pipeline to generate heat for the conference and swim facilities. Another was the pumping of reclaimed water from five miles away into the more than 14 million gallons of storage space on site to irrigate the golf courses and landscaped areas. This 1980 photograph is of the hotel tower during construction. (COI.)

On the western edge of the Industry Hills property was built the Industry Hills Equestrian Center, providing stables for 300 horses, training and show arenas, and horse-riding lessons and rentals. From the very first stages of planning for a city park, the City of Industry included the setting aside of acreage in its natural state for the public to enjoy. To this end, a network of six miles of trails was developed through the hillsides of brush and trees that served as horseback riding and walking paths as well as a cross country course for local high schools. Above is c. 1981 aerial view of the early version of the center, and below is a c. 1982 photograph of unidentified riders on the equestrian trail. (Both, COI.)

The development of the Puente Hills Mall was a large step in diversifying the city's industrial base with a commercial component. The next phase of the commercial expansion was the development of 141 acres at the east end of the mall that was named Puente Hills East. The c. 1978 aerial photograph, taken looking west toward the mall, shows a combination of restaurants, auto dealerships, and entertainment venues being built or completed. The Showboat entertainment center is at the bottom right, as well as Skate Country just to its left. Other buildings, including office space, were incorporated into this initial Puente Hills East construction. The c. 1975 photograph of the Puente 6 Theatre was the first of many large screen complexes that were eventually built in the area. (Both, COI.)

As part of the city's general plan, and using the redevelopment process, was the creation of the City of Industry Civic-Financial Center. The financial component had a head start as Bank of America, United California Bank, and Community Bank already had branches within the site. Shown here at the 1975 groundbreaking for Bank of America's new 25,000-square-foot office and branch building are, from left to right, James Stafford (C.C. Stafford Mill), Robert Rope (city manager), Graham Ritchie (redevelopment agency), Edward Roski (Majestic Realty), Scella Harrison (redevelopment agency), Marty Elenbaas (Bank of America), Annie Faure (redevelopment agency), Art Thomas (Art Thomas Machinery), John Ferrero (mayor), Samuel J. Parriott (council member), Darius Johnson (council member), unidentified, and Charles Rowland (mayor pro tem). The photograph below is a c. 1984 aerial view of the center before the Stafford mill site was razed. (Both, COI.)

As the city continued to grow, so did its need for a larger, more modern city hall. The photograph above shows the original building, built in 1958 on Hacienda Boulevard. The below photograph is of the second city hall, located north of the original building and on Stafford Street (Mayor Dave Way), built in 1975 in the City of Industry Civic-Financial Center. Behind the newer building is the Warehouse Food Outlet and the downtown area of La Puente. In the foreground is the vacant site of the current city hall. (Both, COI.)

Eusebia Salvadora Ebilcieta was born in 1891 in Roncal, Spain. Following her older sister to the United States, she arrived in New York in 1911, just eight days before her 20th birthday. Soon after, she was working as a cook on the same ranch as her sister in Brea. In 1913, she married Matias Mayo, and they had two children, Lawrence and Christine. Tragically, Matias died in 1919 at the age of 26, so Eusebia went back to work at a ranch in Walnut. She married Serafin Sotro in 1922, and they moved to Puente where Serafin, and eventually Lawrence, worked at the Stafford Feed Mill. "Mama" Sotro was known for her volunteer work and Basque style of cooking for church and social gatherings. She is shown here in 1977 at the opening of Sotro Street, named in her honor. From left to right are Samuel J. Parriott (council member), Monsignor Thomas Kiefer, Charles Rowland (council member), John Ferrero (mayor), Sotro, John Van de Kamp (district attorney), and Bill Campbell (state senator). (COI.)

Part of the city's general plan was the inclusion of a historical element. The Workman and Temple Homestead was a natural candidate to fulfill this component of the unique municipality. While the purchase of the Workman House and El Campo Santo Cemetery was completed in 1963, La Casa Nueva was not bought until 1975. Extensive restoration of the buildings and site were completed in 1981 at a cost of $3 million. Since its opening, the Workman and Temple Family Homestead Museum has received awards and recognition not only for the restoration of the site, but for the programs offered to the public. Local fourth-grade students participate in *A Journey Through Time*, for which the city provides the funds for the program and transportation. When this c. 1985 aerial view is compared to the photograph taken in 1921 (shown in chapter 1), there are many obvious changes that have taken place in the area. Among the constants in the two images are the reconfigured San Jose Creek, the Workman House, and the track of the Union Pacific Railroad. (COI.)

Two men in this group photograph, taken in 1986, were instrumental in guiding the city through the booming development period of the 1970s and 1980s—John Radecki (second from left) and Christopher Rope (far right). Radecki was hired by the city's civil engineering firm, National Engineering, in 1963 and worked on large projects, including the Puente Hills Mall, Industry Hills, and the civic-financial center. He was appointed city engineer in 1977 and redevelopment agency director in 1985, serving until his untimely passing in 1990. Rope was the son of the first city manager, Robert Rope, and brought 20 years of experience in public administration when he was hired in 1982 as assistant city manager and then served as city manager from 1985 until 1998. Pictured are, from left to right, Patrick Perez (council member), Radecki, Al Faure (council member), Manuel Garcia (council member), John Ferrero (mayor), Scella Harrison (redevelopment agency), John Paul Ferrero (council member), and Rope. (COI.)

By 1980, the city's boundaries were being filled in by large industrial buildings that easily identify this municipality from the air, as shown in this photograph from that year. No longer defined by just the railroad lines, the city is surrounded by the bedroom communities whose residents often were the workers who filled its business offices, warehouses, and assembly floors. The population of the area, which county planners in the mid-1950s had estimated would ultimately reach one million residents, was about halfway to that prediction. During the 1970s, the city saw an explosion of industrial growth as the number of industrial, manufacturing, and distribution firms went from fewer than 400 to over 900. Commensurate with that increase was the number of employed in the city, which went from 30,000 to 60,000. Yet, there was still approximately 45 percent of the city's 10 square miles of land undeveloped. (COI.)

These two photographs, both taken about 1984, depict the John A. Rowland home. When it was built in 1855, it was the first house made of brick in the region. In 1898, improvements were made to the house by Rowland's daughter and son-in-law, Victoria and Josiah W. Hudson, the house owners and occupants. It was passed on to their daughter Lillian, who married William H. Dibble, and lived in the home until her death in 1959. The house today is located in the City of Industry and in the possession of the La Puente Valley Historical Society. (Both, WTFHM.)

In 1976, the city purchased El Encanto Healthcare and Habilitation Center from the Brown family. This transaction and the oversight role that the city took on was to ensure the continuation of quality specialized healthcare and the safety and well-being of community residents. The skilled nursing facility is well regarded within the healthcare industry for its quality of service, programs, and accommodations. This 1980s aerial view shows the location of the quadrangle facility toward the lower right and adjacent to the Workman and Temple Family Homestead Museum with its well-landscaped grounds that serve as an outdoor site for the clients of El Encanto to visit regularly, as has been done since the 1940s. The city's general plan called for a full-service hospital that would have been located on the vacant lot in the center and lower left of the photograph; however, that plan never came to fruition. (WTFHM.)

In 1986, a group consisting of businesspeople, community leaders, and service organizations started the Industry Hills Charity Rodeo held in the main arena of the equestrian center. That first rodeo was an amateur event and raised $3,500 for the Delhaven Community Center. After that first rodeo, the Professional Rodeo Cowboys Association sanctioned the event and began attracting the top cowboys on the professional circuit. With the support of the City of Industry, its business community and thousands of volunteers, the rodeo has to date contributed $2.4 million to local charities that support disadvantaged youth. These photographs are of the barrel racing competition in the inaugural 1986 rodeo and of the flag presentation ceremony around 1990. (Both, courtesy of the Industry Hills Charity Pro Rodeo.)

As part of the Industry Urban Development Agency Project Area No. 2, the eastern third of the city was mostly undeveloped when this aerial image was taken in 1989. This view is looking toward the southeast, and the graded path of the Grand Avenue extension project can be seen at the top. When Grand Avenue was opened in 1991, it provided access to 600 acres of land for some of the last commercial and industrial development in the city. Shown at the Grand Avenue opening ceremony are, from left to right, Lu King (IMC), Dick Register (IMC), Paul Horcher (assemblyman), Esteban Torres (congressman), Michael Antonovich (county supervisor), John Ferrero (mayor), Trudy Marino, (IMC), W.W. Steiner (IMC), Jay Kim (Diamond Bar mayor), and unidentified. (Both, COI.)

In 1991, local parks were rife with gang activity, rendering them unsafe to the youth in nearby communities. The Industry Sheriff's Station partnered with the City of Industry and its businesses to create the Youth Athletic League (YAL). The YAL provides sports leagues free of charge throughout the year with a deputy on-site. Other activities are camping trips with challenge courses and team-building exercises, scuba diving, and martial arts. In the first year alone, the city contributed $85,000. The photograph above is of a group of local kids participating at an outing to Camp Courage, and the photograph below is of a check presentation in 1991. Attendees are, from left to right, unidentified, Capt. Mike Nagaoka (founding member), Dave Perez (founding member), Tom Moreland (Gabriel Foundation), Ken Rammell (Gabriel Foundation), Deputy Don Alexander, and Deputy Lance Little. (Both, YAL.)

A study done in 1982 showed that, of the city's 54,000 jobs, those related to manufacturing numbered 45,000, and large lots were still available and affordable for new and expanding plants. But, by the late 1980s, manufacturers were leaving California in waves due to increased government regulations and taxes, defense contract cuts, and rising costs of living. By the mid-1990s, a growing commercial presence in the city, loosening of some state regulations, and a foreign trade trend helped stem the tide of the flight of businesses to other states and Mexico. At that time, there were business trade groups coming to the city from China and Taiwan that spurred a large Pacific Rim presence in the city with the manufacturing of food and computers as well as import/export operations. Pictured here in 2003 at the grand opening of Acme Furniture are, from left to right, unidentified, Ron Whittemore (IMC), Ron McPeak (IMC), Dave Perez (mayor), unidentified, Mrs. Chen, George Chen (Acme president), and two unidentified persons. (COI.)

As one solution to the ever-increasing traffic congestion on area freeways, the Metropolitan Transit Authority created the Metrolink commuter rail system. The Riverside line opened in 1994 with a station at Brea Canyon and Currier Roads in the City of Industry and is shown here, around 2000. The property was part of the redevelopment agency with the city as owner of the station and responsible for the security and maintenance of the site. At one time, the city sponsored an electric vehicle project that made several electric cars available at no cost for commuters to use to and from the station. In 2013, a 600-space park and ride structure of Foothill Transit opened on the site for regional bus service where the city provides the land lease for a dollar per year. (COI.)

Alvan Tyler Currier was born in Maine in 1840. He came to this area in 1869 and purchased 2,500 acres in what is now the east end of the City of Industry. Currier was considered a leader in the early development of California agriculture, raising hay, grain, cattle, and fruit on his land. He became a well-known civic and philanthropic figure in the area and was elected Los Angeles County Sheriff in 1881 and state senator in 1898. Currier died at his home on the ranch in 1921. The Currier mansion is shown here, around 2000, among the palm trees on what was part of his ranch. Before development of the Grand Crossing area, the City of Industry donated the mansion to the Historical Society of the Pomona Valley and covered the cost of moving the historic structure down Valley Boulevard to the Phillips Mansion historic site. (COI.)

The Roy Benton Feed Yard was located on Valley Boulevard east of Grand Avenue since 1932 and is shown here in 1998 shortly before its demolition. Previously, Benton worked for the Pacific Cotton Oil Company in Chula Vista, where they used the leftover cottonseed to form a cake on which cattle thrived. Joining his operation in what became the City of Industry was his son Alfred (Al), a two-time recipient of the Distinguished Flying Cross and who also earned a Silver Star for valor during World War II as a fighter pilot. Before the war, Al graduated from the University of California at Davis with a bachelor of sciences degree in animal husbandry. He became the sole operator of the 80-acre feed yard when Roy died in 1959. Al was also involved in the California Cattlemen's Association and the University of California at Davis Livestock Research Advisory Committee and was selected in 1963 to be one of the city's first planning commissioners. He lived to be 100 years old, passing away in December 2020. (COI.)

In 1998, the San Gabriel Valley Council of Governments (COG) created the Alameda Corridor East-Project (ACE) to separate railroad crossings from automobile, truck, and pedestrian traffic. Council member Jack Phillips was the city's representative on the COG's board of directors and fostered a loan from the city to ACE for seed money. This led to improvements at 58 crossings throughout the valley at an estimated cost of $1.7 billion for the railroads that carry 16 percent of the nation's oceangoing containers. Nine of the improved crossings are in the city. These photographs show the construction of the grade separation at Seventh Avenue in 1998 and council member Phillips after the project's completion in 2000. (Above, courtesy of CNC Engineering; below, COI.)

One of the most amazing people to be associated with the City of Industry was the city's treasurer of 52 years, Phyllis Ferrero Tucker. Known for her sharp wit, attention to detail, and stylish wardrobe, Phyllis dedicated herself to the city up until her death in 2017 at 94 years of age. She was born in 1922 in the family barn on Proctor Avenue, as their house was being constructed at the time. Like her brother, mayor John, Phyllis's work ethic likely came from her mother, Teresa. Having lost her husband in 1930, Teresa worked the family farm with John, and was employed at the Sunkist Packing Company until their land was owned free and clear, and the children were educated and married. Once the Ferrero home was built, Phyllis never moved out, living there her entire life and in the city that she loved. (COI.)

The freeways that were built back in the 1960s and 1970s were having a hard time keeping up with the amount of traffic in the 21st century. The design of the confluence between the Pomona and Orange freeways was adequate for the time it was built; however, it could not handle the increases in population and truck traffic with the transportation of goods to and from the ports of Los Angeles and Long Beach decades later. In a partnership with federal, county, and state agencies that harkens back to the Azusa Avenue overpass project in 1967, the 57/60 Confluence Project saw the city commit $35 million to improve the nation's sixth worst segment of freeway. Photographed here at the opening of a new on-ramp in 2017 are, from left to right, Cathy Marcucci (council member), Cory Moss (mayor pro tem), Mark Radecki (mayor), Jimmy Lin (Diamond Bar council member), and Steve Tye (Diamond Bar council member). (COI.)

In 1978, the city purchased the 2,445-acre Tres Hermanos Ranch, which lies to the east and southeast in the cities of Diamond Bar and Chino Hills. City leaders envisioned a potable water reservoir for much of the property; however, that and other plans were ultimately left unpursued. In 2019, Industry's city council entered into a joint powers agreement with Diamond Bar and Chino Hills to maintain the land as open space. The name of the ranch comes from previous owners and partners in the oil business, Harry Chandler, William Rowland, and William B. Scott. This 1925 photograph shows Scott's daughter Josephine astride a horse on the pasture that cattle have grazed since the Mexican era of the ranchos, and it appears that it will remain much as it was during that time and of the period of Workman and Rowland on the neighboring Rancho La Puente. (WTFHM.)

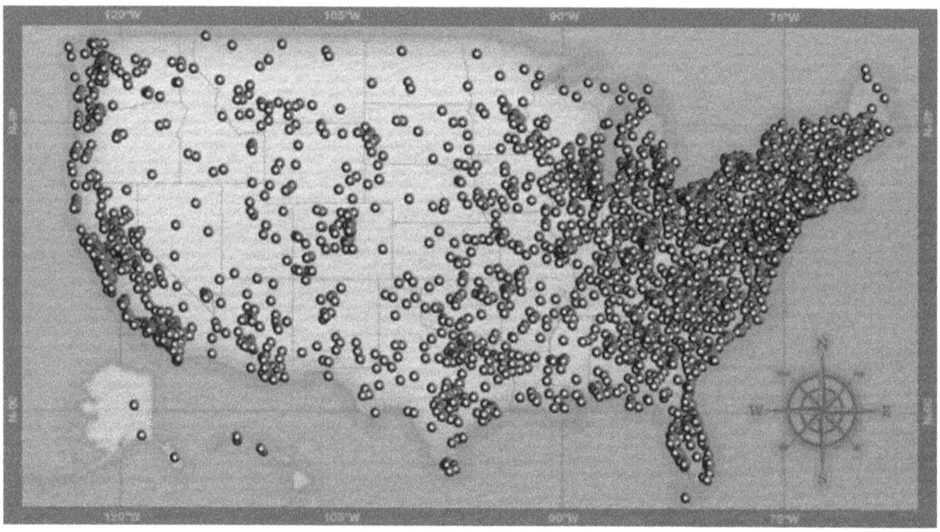

www.ingramcontent.com/pod-product-compliance
Lightning Source LLC
LaVergne TN
LVHW071100050926
847404LV00038B/472
9781467107785